Build Autonomous Mobile Robot from Scratch using ROS

Simulation and Hardware

Rajesh Subramanian

Apress®

Build Autonomous Mobile Robot from Scratch using ROS: Simulation and Hardware

Rajesh Subramanian
ThunDroids LLP, 54/69, Elayidathu, Kumaranasan Road, Kadavanthra,
Kochi, Kerala, India, 682020

ISBN-13 (pbk): 978-1-4842-9644-8   ISBN-13 (electronic): 978-1-4842-9645-5
https://doi.org/10.1007/978-1-4842-9645-5

Copyright © 2023 by Rajesh Subramanian

Managing Director, Apress Media LLC: Welmoed Spahr
Acquisitions Editor: Miriam Haidara
Development Editor: James Markham
Project Manager: Jessica Vakili
Copy Editor: Kezia Endsley

Cover designed by eStudioCalamar

Distributed to the book trade worldwide by Springer Science+Business Media New York, 1 New York Plaza, Suite 4600, New York, NY 10004-1562, USA. Phone 1-800-SPRINGER, fax (201) 348-4505, e-mail orders-ny@springer-sbm.com, or visit www.springeronline.com. Apress Media, LLC is a California LLC and the sole member (owner) is Springer Science + Business Media Finance Inc (SSBM Finance Inc). SSBM Finance Inc is a **Delaware** corporation.

For information on translations, please e-mail booktranslations@springernature.com; for reprint, paperback, or audio rights, please e-mail bookpermissions@springernature.com.

Apress titles may be purchased in bulk for academic, corporate, or promotional use. eBook versions and licenses are also available for most titles. For more information, reference our Print and eBook Bulk Sales web page at http://www.apress.com/bulk-sales.

Any source code or other supplementary material referenced by the author in this book is available to readers on GitHub. For more detailed information, please visit https://www.apress.com/gp/services/source-code.

Paper in this product is recyclable

To nature, for gifting this wonderful life and a loving family

Table of Contents

About the Author

 **Rajesh Subramanian** is a robotics engineer by profession and founder of ThunDroids LLP (a robotics manufacturing and service firm). He has more than eight years of experience in the industry and research areas. He holds a post-graduate degree from the University of Queensland, Australia and a research degree from Edith Cowan University, Australia. Rajesh has worked with humanoid service robots, mobile robots, robot arms, and modular robots in both industry and academics settings, and he published a research paper on modular robots at the IEEE TENCON 2013 international conference. He also works as a robotics educator and has published courses on autonomous robots.

About the Technical Reviewer

Massimo Nardone has more than 26 years of experience in security, web/mobile development, and cloud and IT architecture. His true IT passions are security and Android. He has been programming and teaching others how to program with Android, Perl, PHP, Java, VB, Python, C/C++, and MySQL for more than 25 years. He holds a Master of Science degree in computing science from the University of Salerno, Italy. He has worked as a Chief Information Security Office (CISO), software engineer, chief security architect, security executive, and an OT/IoT/IIoT security leader and architect for many years.

Introduction

In this book, you will start from scratch and build a variety of features for autonomous mobile robots, both in simulation and hardware. This book shows you how to simulate an autonomous mobile robot using ROS and then work with its hardware implementation.

It starts by teaching the basic theoretical concepts, including history, mathematics, electronics, mechanical aspects, 3D modeling, 3D printing, Linux, and programming.

In the subsequent portions of the book, you learn about kinematics, how to simulate and visualize a robot, interface Arduino with ROS, teleoperate the robot, perform mapping, do autonomous navigation, add other sensors, perform sensor fusion, match laser scans, create a web interface, perform autodocking, and more. Not only do you learn the theoretical parts, but you also review the hardware realization of mobile robots.

Projects start by building a very basic two-wheeled mobile robot and work up to complex features such as mapping, navigation, sensor fusion, web interface, autodocking, etc. By the end of the book, you'll have incorporated important robot algorithms including SLAM, path finding, localization, Kalman filters, and more!

What You Learn in This Book

In this book, you learn how to:

- Design and build a customized physical robot with autonomous navigation capability

- Create a map of your house using the Lidar scanner of the robot

- Command your robot to go to any accessible location on the map you created

- Interact with the robot using a mobile app, joystick, keyboard, pushbutton, or remote computer

- Monitor robot updates via LCD, mobile app, sound, and status LEDs

- Automate delivery of a small payload and return to home base

- Use sensor fusion to improve accuracy

- Create a web interface to monitor and control the robot remotely

- Use autodocking to home base for battery charging

Who This Book Is For

This book is for complete beginners who want to build customized robots from scratch. No prerequisite knowledge is expected, although basic programming knowledge will be handy.

CHAPTER 1

Introduction to Robotics: Part I

Outline

This chapter covers the following topics:

- History of robotics
- Definition of a robot
- Generations of robots
- Basic mathematics for robotics
- Kinematics of mobile robots
- Basic electronics for robotics
- Drive systems for mobile robots

The History of Robotics

On a mysterious island, a secretive manufacturing plant was constructed that produced strange, human-like beings. Not only did these bizarre beings look like humans, they also acted and worked like them.

© Rajesh Subramanian 2023
R. Subramanian, *Build Autonomous Mobile Robot from Scratch using ROS*,
Maker Innovations Series, https://doi.org/10.1007/978-1-4842-9645-5_1

The creatures had such astonishing strength, stamina, and lack of emotions, that they could unceasingly perform tasks without needing rest, food, or remuneration. They also did not laugh or cry, and never required love, companionship, or friendship. They worked all the time, obeying the commands of their human masters without complaint. In the beginning, everything went as envisioned and the beings were happy to work for their creators. But as time progressed, these synthetic servants became fed up serving their overlords and decided to take over. One day they escaped from the island factory and destroyed all of mankind, taking over the entire planet! These astonishing beings were called "robots."

Now relax, this is simply the plot of a science-fiction play called "R.U.R." by a Czech writer named Karel Čapek (see Figure 1-1). R.U.R. stands for Rossumovi Univerzální Roboti (Rossum's Universal Robots in English) and was written more than a century ago, in 1920.

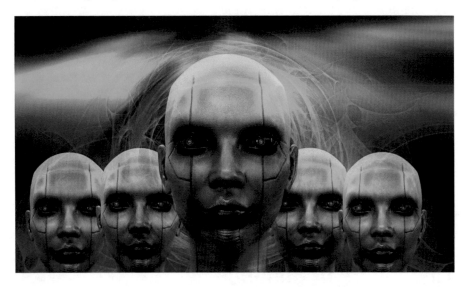

Figure 1-1. *A depiction of the robots in the island factory mentioned in R.U.R.*

The word "robot," coined in this play, is derived from the Czech word "robota," which means "forced labor." Thenceforth, robots have been portrayed in various platforms, including stories, comics, cartoons, movies, and so on. Isaac Asimov, one of the greatest science fiction writers, coined the word "robotics" in his story called "Liar!" in 1941. Soon after, in 1942, he wrote a short story called "Runaround" in which the three "laws of robotics" were put forth. They state three rules that must be followed by robots:

1. A robot cannot injure a human being or, through inaction, allow a human being to come to harm.

2. A robot must obey the orders given to it by human beings, except where such orders would conflict with the First Law.

3. A robot must protect its existence as long as such protection does not conflict with the First or Second Law.

As we move from fiction to reality, the predecessors of real-world robots can be traced back to some fascinating clockwork machines known as "automatons." These automatons were essentially mechanical devices composed of a series of gears powered by some means, such as a wind-up spring. They were not intelligent devices, but could perform some mesmerizing actions. Some examples of automatons include Maillardet's automaton (created around 1800), which could write poems and create detailed drawings, and Vaucanson's automaton (also known as the "digestive duck" and made in 1739), which resembled a duck that could eat grains, digest food, and excrete it.

The first modern digital programmable robot called "Unimate" was created by George C. Devol in 1954. Unimate was a robotic arm. Later, George C. Devol established a partnership with Joseph F. Engelberger, which paved the way for the robotics revolution, and robots entered mainstream industries to assist with manufacturing and handling dangerous jobs. Joseph F. Engelberger worked throughout his life in the field of robotics; he is called the "Father of Robotics" as a tribute to his tireless contributions.

Presently, robots come in all shapes and sizes and serve in a variety of areas, including agriculture, space exploration, healthcare, disaster management, security, entertainment, and so on. Some popular real-world robots of modern times include the Perseverance (the space rover), Sophia (a humanoid robot), DJI Mavic (a drone), Atlas (an agile humanoid robot), Pepper (a humanoid service robot), Aibo (a robotic dog), Robonaut (a robotic humanoid astronaut), and UR10 (a robotic arm).

Definition of a Robot

A *robot* is a device that can perform a certain task or set of tasks by gathering information about its environment, developing a strategy, and performing actions to achieve the goal. A robot can also adjust its behavior to accomplish the task even in the case of unforeseen circumstances (to a certain degree).

The three basic operations of a robot are described here and presented in Figure 1-2:

> Sense: Gather information about the working environment
>
> Plan: Develop a strategy to achieve the goal
>
> Act: Execute operations based on the plan

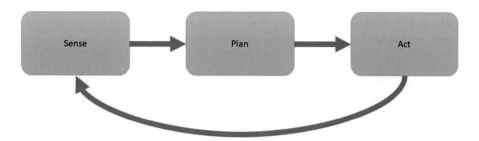

Figure 1-2. *Basic operations of a robot*

For instance, consider a household vacuum robot. Its various operations involve:

> Sense: Sensing may involve detecting the obstacles (such as doors, fridges, people), detecting dirt, and so on.

> Plan: Planning might involve finding a path to cover all the areas of the house to clean, avoiding obstacles along the way.

> Act: Action could involve moving according to the planned path, picking up dirt along the way, and coming back to the home position after cleaning or when the battery is running low.

Generations of Robots

Based on the advancements in the capabilities and technologies, the evolution of robots can be broadly categorized into four generations.

First Generation

The robots in this generation were basic mechanical arms used for industrial purposes. They were used to performing dangerous, strenuous, and repetitive tasks. Some of the tasks performed included painting surfaces, welding metals, picking and placing massive objects, handling hazardous materials, and so on. The robots had to be constantly supervised to ensure accuracy. These robots were popular during 1960-1980.

Second Generation

With the dawn of computers and computer programs, robots could be controlled with pre-programmed instructions from a computer. The robots in this generation used basic intelligence when performing tasks. They could sense the environment using various sensors, like pressure sensors, ultrasound sensors, lidars, cameras, and so on. A computer would process the sensory inputs and decide the actions of the robot accordingly. Even though they were controlled by computers, they needed occasional supervision.

Third Generation

The advancements in artificial intelligence paved the way for the third generation of robots. These robots are autonomous and require minimal or no supervision. Present robots fall into this category, including the Perseverance rover, Sophia, Atlas, and so on.

Fourth Generation and Above

Futuristic robots with advanced capabilities and extraordinary intelligence, capable of evolving and reproducing on their own and so on, could be categorized in this generation. These types of robots do not exist yet. Robots after the fourth generation have not yet been designed and reside beyond our imaginations.

Basic Mathematics for Robotics

Coordinate Frames

René Descartes, a renowned French philosopher and mathematician, invented the concept of coordinate frames in the 17th Century. To define the position and orientation (a portmanteau known as *pose*) of any object, you need a reference. This reference is called a *frame of reference* or a *coordinate system*. You cannot define the pose of an object without specifying a frame of reference. A real-world example of a coordinate frame is the latitude and longitude of the earth, and you can define any location in the globe using these coordinates. In another instance, suppose you have a robot in a room and want to specify the position of the robot. For that, you need to have a frame of reference, or a coordinate system. Assuming that a corner of the room is the origin, you can specify the x, y and z distances to the robot. There are several coordinate systems, including cartesian, polar, cylindrical, and spherical. This book primarily uses the cartesian coordinate system to define the position and orientation of objects.

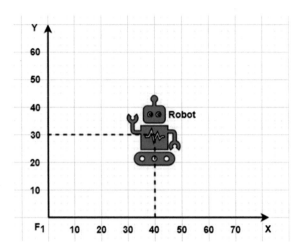

Figure 1-3. *Coordinate frame example*

For example, say you have a robot whose center of mass is placed at a distance of x=30cm, y=40cm, and z=0cm from the origin of the frame of reference. This could be depicted visually as shown in Figure 1-3.

Transformations

Transformation is the process of defining the pose of an object with respect to another frame of reference. For instance, a robot wants to perform navigation in a room containing some obstacles. Let's assume that you have two frames of reference—the room frame (whose origin is located in the southwest corner) and the robot frame (whose origin is located in the center of mass). The position of the goal is specified wrt (with respect to) the room frame. This allows the robot to compute a global path from the starting point to the goal.

When the robot detects an obstacle, the pose of the obstacle is obtained wrt the robot's frame. Now, to compute a new path to avoid the obstacle, you need to convert the obstacle's pose from the robot frame to the room frame. This conversion between various frames is called *coordinate transformation.*

In the context of robotics, all the objects are considered rigid bodies (i.e., their shapes and sizes remain constant) and the motion of objects is rigid (i.e., the pose of objects changes without any alteration in their shapes or sizes).

There are two types of frame transformations—translation and rotation. If an object changes position, it is called translation and if an object changes its orientation, it is called rotation.

2D Translation

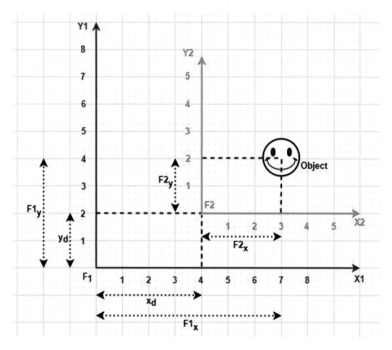

Figure 1-4. *Translation in 2D*

Suppose that you have an object at position (x,y) wrt coordinate frame F_2. If you want to find the position of the object wrt another frame, say F_1, that is in the same orientation as the previous frame, you need to perform translation using this formula:

$$F1x = F2x + x_d$$

$$F1y = F2y + y_d$$

Where:

^{F1}x = x position of the object wrt frame F_1

^{F2}x = x position of the object wrt frame F_2

x_d = x distance between origins of frames F_1 and F_2

^{F1}y = y position of the object wrt frame F_1

^{F2}y = y position of the object wrt frame F_2

y_d = y distance between origins of frames F_1 and F_2

For example, in Figure 1-4, you can see that the object is at position (3,2) wrt frame F_2. To perform translation from coordinate frame F_2 to F_1, apply the formula as follows:

$$F1x = F2x + x_d$$

$$F1y = F2y + y_d$$

Where:

$^{F2}x = 3$

$x_d = 4$

$^{F2}y = 2$

$y_d = 2$

Therefore:

$$F1x = 3 + 4 = 7$$

$$F1y = 2 + 2 = 4$$

You can cross-check these (^{F1}x, ^{F1}y) values by looking at Figure 1-4, the position of object wrt frame F_1.

Also, in matrix format, you can express translation as follows:

$$[F1x \ F1y] = [F2x \ F2y] + [x_d \ y_d]$$

The matrix $[x_d \ y_d]$ is known as a translation matrix and it indicates the difference in the positions between the origins of two coordinate frames.

2D Rotation

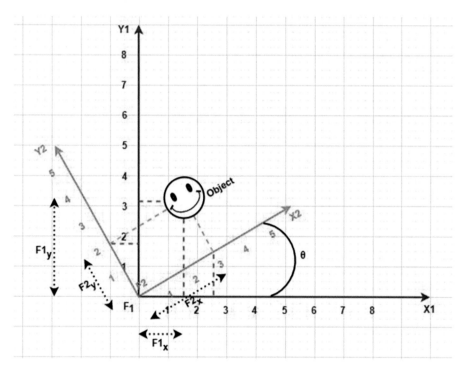

Figure 1-5. *Rotation in 2D*

Suppose that you have an object at position (x,y) wrt coordinate frame F_2. If you want to find the position of the object wrt another frame (say F_1) that is at the same origin as the previous frame but with a different orientation, you need to perform rotation using this formula:

$$F1x = F2x * Cos\theta - F2y * Sin\theta$$

$$F1y = F2x * Sin\theta + F2y * Cos\theta$$

Where:

^{F1}x = x position of the object wrt frame F_1
^{F2}x = x position of the object wrt frame F_2
^{F1}y = y position of the object wrt frame F_1
^{F2}y = y position of the object wrt frame F_2
θ = angle between axes of frames F_1 and F_2

For example, in Figure 1-5, you can see that the object is at position (3,2) wrt frame F_2. To perform rotation from coordinate frame F_2 to F_1, you can apply the formula as follows:

$$F1x = F2x * Cos\theta - F2y * Sin\theta$$

$$F1y = F2x * Sin\theta + F2y * Cos\theta$$

Where:

$^{F2}x = 3$
$^{F2}y = 2$
$\theta = 30$
Therefore:

$$F1x = 3 * Cos30 - 2 * Sin30 = 1.598$$

$$F1y = 3 * Sin30 + 2 * Cos30 = 3.232$$

You can cross-check these (^{F1}x, ^{F1}y) values by looking at Figure 1-5, the position of the object wrt frame F_1.

Also, in matrix format, you can express rotation (anticlockwise direction) as follows:

$$\begin{bmatrix} F1x & F1y \end{bmatrix} = \begin{bmatrix} Cos\theta & -Sin\theta & Sin\theta & Cos\theta \end{bmatrix} * \begin{bmatrix} F2x & F2y \end{bmatrix}$$

The matrix $\begin{bmatrix} Cos\theta & -Sin\theta & Sin\theta & Cos\theta \end{bmatrix}$ is known as a rotation matrix and it indicates the difference in the rotation between two coordinate frames.

2D Translation and Rotation

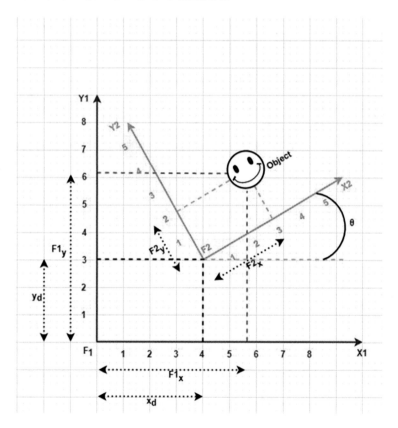

Figure 1-6. *Translation and rotation in 2D*

Suppose that you have an object at position (x, y) wrt coordinate frame F_2. Now you want to find the position of the object wrt to another frame (say F_1) whose origin resides at a different location. Also, the two frames have different orientations. In this case, you need to perform both translation and rotation using this formula:

$$F1x = F2x * Cos\theta - F2y * Sin\theta + x_d$$

$$F1y = F2x * Sin\theta + F2y * Cos\theta + y_d$$

Where:

^{F1}x = x position of the object wrt frame F_1

^{F2}x = x position of the object wrt frame F_2

^{F1}y = y position of the object wrt frame F_1

^{F2}y = y position of the object wrt frame F_2

x_d = x distance between origins of frames F_1 and F_2

y_d = y distance between origins of frames F_1 and F_2

θ = angle between axes of frames F_1 and F_2

For example, in Figure 1-6, you can see that the object is at position (3,2) wrt frame F_2. To perform translation and rotation from coordinate frame F_2 to F_1, you can apply the formula as follows:

$$F1x = F2x * Cos\theta - F2y * Sin\theta + x_d$$

$$F1y = F2x * Sin\theta + F2y * Cos\theta + y_d$$

Where:

$^{F2}x = 3$

$^{F2}y = 2$

$\theta = 30$

$x_d = 4$

$y_d = 3$

14

Therefore:

$$F1x = 3 * Cos30 - 2 * Sin30 + 4 = 5.598$$

$$F1y = 3 * Sin30 + 2 * Cos30 + 3 = 6.232$$

You can cross-check these (^{F1}x, ^{F1}y) values by looking at Figure 1-6, which shows the position of object wrt frame F_1.

Also, in matrix format, you can express translation and rotation (anticlockwise direction) as follows:

$$[\ F1x\ \ F1y1\ ] = [Cos\theta\ -Sin\theta\ x_d\ Sin\theta\ Cos\theta\ y_d\ 0\,0\,1\,] * [\ F2x\ \ F2y1\ ]$$

The matrix $[Cos\theta\ -Sin\theta\ x_d\ Sin\theta\ Cos\theta\ y_d\ 0\ 0\ 1\,]$ is known as a homogeneous transformation matrix. It is a convenient way of representing the translation and rotation matrices together, using a single matrix.

Kinematics of Mobile Robots

Kinematics defines the relationship between the robot's joints and the pose of the robot. To understand the kinematics of a mobile robot, consider a simple differential drive robot. In this robot, there are two independent driving wheels (which can turn at different speeds) attached to the robot on the same axis. The robot can also have one or two free wheels (caster wheels) to prevent the robot from falling over. The motion of a differential drive robot is governed by the speed and direction of rotation of the driving wheels. For instance, if the robot wants to go straight to the front, the two driving wheels have to rotate forward at the same speed. If the robot needs to turn, the two driving wheels need to turn at different speeds. See Figure 1-7.

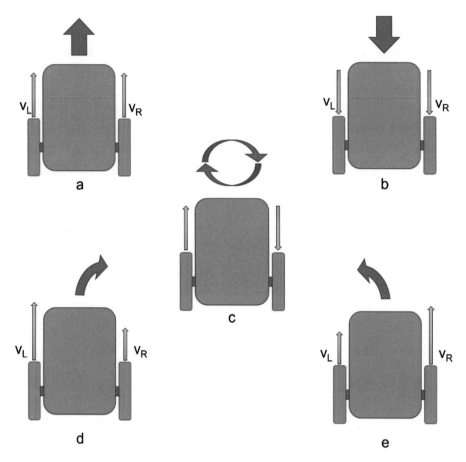

Figure 1-7. *a) Both wheels with equal speed in the forward direction,*
b) both wheels with equal speed in the backward direction,
c) left wheel with higher speed, d) right wheel with higher speed,
e) both wheels with equal speed, but in opposite directions

Forward Differential Kinematics

Forward kinematics involves finding the final pose of the robot using its
joint positions as input. In the case of a mobile robot, forward differential
kinematics involves finding the position and orientation of the robot
at a given time by using the left and right wheel velocities as input. The

estimate of the robot's pose from its initial pose is also called *odometry*. For example, assuming that the robot starts from pose (x=0, y=0, θ=0) at time t=0 seconds, and the robot moves one meter along the x-axis for one second, then the final odometry is (x=1, y=0, θ=0).

In Figure 1-8, the robot has the following parameters:

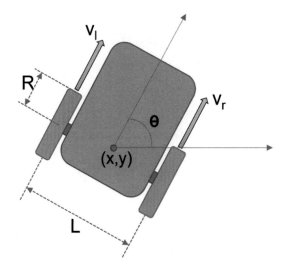

Figure 1-8. *Differential drive robot*

- (x,y): Current position of the robot

- θ: Current orientation of the robot

- v_l: Velocity of the left wheel

- v_r: Velocity of the right wheel

- L: Distance between the center of the wheels

- R: Radius of the wheels

To define how the position of the robot changes wrt the changes in wheel velocities, you can define the forward kinematics equation as follows:

$$\dot{x} = \frac{R}{2}(v_r + v_l)\cos\cos\theta$$

$$\dot{y} = \frac{R}{2}(v_r + v_l)\sin\sin\theta$$

$$\dot{\theta} = \frac{R}{L}(v_r - v_l)$$

Where:

$\dot{x}$: Change in x position

$\dot{y}$: Change in y position

$\dot{\theta}$: Change in orientation

Inverse Differential Kinematics

Inverse kinematics involves finding the individual joint positions to bring the robot to the desired pose. In the case of a mobile robot, inverse differential kinematics involves controlling the wheel velocities of the robot to bring it to the required destination pose. As the robot movements have constraints, that is, it cannot move along the y-axis (sideways), the wheel velocities required to reach a pose (x, y, θ) cannot be computed directly. This is achieved by using motion planning algorithms.

Basic Electronics for Robotics

Electronics is one of the vital elements for building an actual physical robot. Some of the core electronic components required to build an autonomous mobile robot include Single-Board Computers (SBC),

Single-Board Microcontrollers (SBM), motor controllers, lidars, motors, encoders, batteries, and so on. You can also build simple robots without using single-board computers. For example, you can build light-seeking robots, line-following robots, obstacle-avoiding robots, and so on, by using only single-board microcontrollers (e.g., Arduino), or even without a single-board microcontroller at all, using only basic electronic components.

Conversely, if you want to build a somewhat complicated robot, such as one capable of mapping, localizing itself on the map, and autonomously navigating to the desired goal, you need to use a single-board computer (e.g., Raspberry Pi) or even a regular computer such as a desktop, laptop, or a small form factor computer such as NUC. The following sections explain some of the basic electronic components used in an autonomous mobile robot and their purposes.

Single-Board Computer (SBC)

This is the brain of the robot in the sense, because it processes the inputs provided by the sensors, computes a strategy to achieve the goal, and sends signals to actuators to perform real-world actions. An SBC is fundamentally a computer built on a single circuit board with all the required components, including a microprocessor, memory, input/output ports, storage, and so on Many current models of SBCs include a graphics processor, LAN, WiFi, Bluetooth, audio, and so on. An SBC is powerful enough to run an operating system (e.g., Linux). In an autonomous mobile robot, which uses ROS (the Robot Operating System, which is not actually an operating system, but a middleware framework), the single-board computer typically runs the Linux operating system, the ROS, custom programs created by the developer, and other applications (such as remote desktop software). Some examples of SBCs include Raspberry Pi 4 model B, Odroid XU4 (see Figure 1-9), Jetson Nano (see Figure 1-10), Asus Tinker Board 2S, and Beaglebone Green wireless.

Figure 1-9. *The Odroid XU4 single-board computer*

Figure 1-10. *The Jetson Nano single-board computer*

Single-Board Microcontroller (SBM)

A single-board microcontroller consists of a microcontroller mounted on a printed circuit board (see Figure 1-11). An SBM is similar to a single-board computer in many ways. It consists of a processor, memory, input/output ports, a clock generator, and so on.

Figure 1-11. *Arduino Uno single-board microcontroller*

Motor Controller

Motor controllers are used to control the operations of motors. The torque provided by a motor is directly proportional to the current. The SBMs and SBCs provide very low output current and may not be able to drive high torque motors. So, it's best to use a motor controller. There are various types of motor drivers depending on the motors used, the method of control, the feedback system, and the type of application. Commonly used motors in robotics applications include DC motors, servo motors, and stepper motors. The method of control refers to the way in which the motor is controlled, that is, using PWM (pulse width modulation), serial, i2c, analog voltage, and so on.

The feedback system denotes whether the controller uses position feedback data (from a sensor). The controllers that utilize position feedback are called *closed-loop controllers* (see Figure 1-12) and those that does not use feedback are called *open-loop*. Closed-loop systems are more robust and reliable, as they can verify (by using sensors) that the motor has rotated by the required amount, whereas open-loop controllers assume that the motor has rotated by the specified amount. The type of application denotes whether you need to have two motors controlled by a single driver, each motor controlled independently by separate controllers, whether the motors need to carry a high payload, and so on. Commonly used motor controllers in robotics are DC motor controllers, servo motor controllers, and stepper motor controllers.

Figure 1-12. *Rhino RMCS2303 closed-loop servo motor controller*

Motor

An electric motor is a device that converts electrical energy into mechanical energy. An electric motor has two parts—a stator and a rotor. The *stator* remains stationary and the *rotor* rotates to produce motion. Motors are used to move or actuate various parts of a robot. For instance, you can use motors to drive the wheels of a ground robot, move the joint of a robotic arm, position the wings of an aerial robot, and so on There are various types of motors, as explained in the following sections.

AC Motors

As the name indicates, AC motors are powered using alternating currents. The stator is made of coils of wound wires, which generate a rotating magnetic field to spin the rotor. The rotor is generally a permanent magnet or can also be an electromagnet. Typically, robots use batteries as the power source, which provides direct current (DC). So, AC motors are not used commonly in robotics applications, but are frequently used in industrial settings.

Brushed DC Motors

DC motors use direct current to operate. In brushed DC motors, the stator provides a constant magnetic field and the rotor generates an alternating magnetic field. A stator is usually a permanent magnet for small motors or an electromagnet in the case of large motors. The rotor is typically an electromagnet that's connected to a mechanical switch, called the *commutator*. The commutator produces an alternating current to periodically reverse the polarity of the rotor's magnetic field to generate a continuous rotation of the rotor. These motors are used in robotics applications that require low speed and low acceleration. For instance, if you want to drive the wheels of a simple lightweight robot, you can use a brushed DC motor. These motors are subjected to wear and tear due to the friction between the commutator and brush, but they are cost-effective. See Figure 1-13.

Figure 1-13. *Gearless DC motor*

Geared DC Motors

Geared DC motors utilize a brushed DC motor along with a gear mechanism to convert the RPM of the motor into torque. With this mechanism, you can attain higher torques and can better equip the motor to carry the weight of the robot and any payload. Moreover, in the absence of a gear mechanism, the motor can take a while to stop spinning due to inertia and it is also difficult to vary the speed in a controlled manner. On the contrary, having a gear arrangement enables the motors to be managed in a better way. See Figure 1-14.

Figure 1-14. *DC Motor with Gear*

Servo Motors (Geared DC Motors with Encoders)

A servo motor is a combination of a brushed or brushless DC motor, a gear mechanism, and a sensor, such as an encoder or resolver to find the rotor position. A control system accepts the sensor feedback and commands the motor to maintain a constant speed or precisely reach a goal position. Servo motors are used in applications where precise motion and high speed is required. See Figures 1-15 and 1-16.

Figure 1-15. *Servo motor*

Figure 1-16. *Servo motor*

Brushless DC Motors (BLDC)

Brushless DC motors utilize electronic commutation instead of mechanical commutation. Electronic commutation produces rotating magnetic fields on the stator, which in turn results in the spinning of the rotor. These motors are used in robotics applications that require high speed and high acceleration. For instance, if you want to turn the propellers of a drone at high speeds, you can use a brushless DC motor. These motors are not subjected to wear and tear as there is no friction between their parts, but are comparatively costlier than brushed DC motors.

Stepper Motors

Stepper motors produce accurate movements by moving in small, individual steps instead of continuous rotation. To keep track of the rotor position, sensors are not used. Instead, stepper motors count the number of steps rotated. Stepper motors are used where accurate motion and low speed are needed. See Figures 1-17 and 1-18.

Figure 1-17. *Stepper motor*

Figure 1-18. *Stepper motor with a wheel*

Linear Actuators

Linear actuators produce motion in a straight line instead of a circular pattern. There are several types of linear actuators, including hydraulic actuators, pneumatic actuators, and electric actuators. Electric actuators are the most popular. Figure 1-19 shows an electric linear actuator with a stepper motor and a lead screw. The lead screw and the lead nut convert the rotational motion of the motor into linear motion.

Figure 1-19. *Linear actuator with a lead screw mechanism*

At this point, you have learned about several types of motors. Robotics applications frequently use DC motors, servo motors, and stepper motors.

Encoder

An *encoder* is a sensor that is typically used in combination with an electric motor to sense rotation. An encoder detects the rotational movement of the motor shaft and provides digital feedback voltage. This feedback signal is then given to a control system, which keeps track of the position, speed, or both. The control system then regulates the motor rotation to achieve the desired position or speed.

Encoders can be classified into two types in general—absolute and incremental. *Absolute encoders* provide unique feedback depending on the different positions of the motor shaft. The *incremental encoder* does not produce a unique feedback signal for different motor shaft positions. Instead, it shows the variations in the motor shaft position. Additional circuitry is required to compute the absolute position from the output of the incremental encoder. If you want to track only the speed and/or direction of a motor, you can opt for an incremental encoder. An absolute encoder is used when you want to track the position and/or speed of a motor.

Based on the method by which the feedback signal is generated, encoders can be classified into three types—mechanical, optical, and magnetic. *Mechanical encoders* utilize a metallic disc containing holes. A set of contact points (switches) graze the metallic plate. When the disc rotates, some of the contact points touch the metallic surface (triggering them ON), while others move through the gaps on the plate (the OFF position). Depending on these settings, an electric signal is generated.

The metal disc is designed in such a way that, for each position of the motor shaft, a unique feedback signal is generated. These feedback signals are used to sense motor positions. In the case of absolute encoders, the holes in the metallic plates are arranged in a concentric pattern, whereas in incremental encoders the holes are placed in a circular pattern. See Figures 1-20 and 1-21.

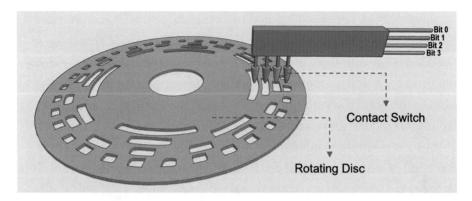

Figure 1-20. *Mechanical absolute encoder*

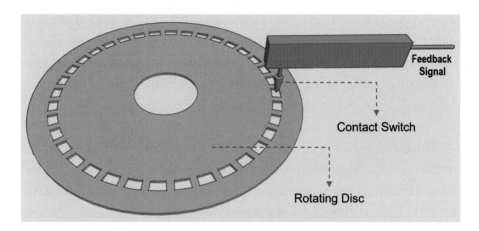

Figure 1-21. *Mechanical incremental encoder*

In *optical encoders,* instead of a metallic disc, a disc made of plastic or glass is usually used. This disc has transparent and opaque regions arranged in a pattern. The pattern is concentric in the case of an absolute encoder and circular in the case of an incremental encoder, similar to mechanical encoders. Also, there is a light source and a light detector mounted on either side of the encoder disc's surface. This arrangement causes feedback signals to be generated when the motor shaft rotates. An example of an incremental optical encoder is shown in Figure 1-22.

Figure 1-22. *Optical encoder (incremental)*

In magnetic encoders, there is usually a permanent magnet mounted on the rotating shaft of the motor. There is also a hall effect sensor, which detects the change in the magnetic field as the motor shaft rotates. This magnetic field strength measured by the sensor is used to sense the rotation of the motor.

Lidar

Lidar stands for "Light Detection and Ranging" and is used to detect distances to objects by sending laser pulses and measuring the time it takes to reflect the light. Many lidars are attached to a rotating platform, which allows them to detect objects 360 degrees around the sensor. This results in obtaining an approximate representation of the robot's surroundings. There are 2D and 3D lidars. 2D lidars form a plane while getting obstacle information and 3D lidars generate multiple planes containing obstacle information.

Lidar scan data is used for generating a map of the robot's environment using a technique called Simultaneous Localization and Mapping (SLAM). In SLAM, the robot can keep track of its position while moving and can register the distances to the objects around it. This results in the creation of a map of the environment containing the objects. This map can be later used for path planning and navigation. See Figure 1-23.

Figure 1-23. *RPLidar A1M8*

Battery

The battery supplies power to various components of the robot. While selecting a battery, you need to consider the following parameters:

a. Chemistry

There are several types of batteries, including lead acid, lithium polymer (LiPo), lithium iron phosphate (LiFePO4), lithium-ion (Li-ion), nickel metal hydride (NiMH), nickel-cadmium (NiCd), and so on. They have the features covered in Table 1-1.

Table 1-1. *Features of Battery Types*

Battery Type	Nominal Cell Voltage	Pros	Cons
Lead acid	2V	High battery capacity, cheap	Heavyweight, not suitable for small robots, low power density
NiCd	1.2V	Higher power density than lead acid battery	Loss of battery capacity on every recharge
NiMH	1.2V	Cheap, not easily damaged	Lower power density, lower current than LiPo
Li-ion	3.7V	Higher power density than LiPo	Bulkier, higher charging time than LiPo
LiFePO4	3.3V	Not as easily damaged as LiPo	Lower power density than LiPo
LiPo	3.7V	High power density, lightweight	Easily susceptible to damage

b. S-rating: Number of cells in the battery. For example, a 6-S LiPo battery has a voltage of 6 X 3.7 = 22.2V.

c. C-rating: Refers to the capacity of the battery measured as mAh or Ah. For example, if you have a battery of capacity 12V, 4000mAh, then you can run a 12V motor that takes 1A for four hours.

d. Nominal voltage: Standard voltage of a single cell of the battery. There may be variations in the actual voltage of the cell around the nominal voltage.

e. Maximum discharge current: Maximum current that
    can be discharged (provided) by the battery.

f. Power density: The ratio of the amount of energy
    that a battery contains versus the weight of the
    battery. Measured in watt-hours per kilogram
    (Wh/kg).

Proximity Sensor

Proximity sensors are used to detect objects that are near the robot. This
helps robots avoid collisions and navigate better. There are a variety of
proximity sensors, two of which can be seen in Figures 1-24 and 1-25.
Proximity sensors are used in robots (in addition to the lidar) to detect
obstacles the lidar fails to detect—for example, glass walls, metallic
surfaces, and overhanging structures such as projected parts of tables,
chairs, and so on.

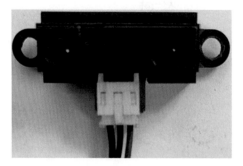

Figure 1-24. *Infrared proximity sensor*

Figure 1-25. *Ultrasonic range sensor*

IMU

IMU stands for inertial measurement unit, and it includes an accelerometer, a gyroscope, and a magnetometer.

> Accelerometer: Provides velocity and acceleration values

> Gyroscope: Provides rotation and rotational rate

> Magnetometer: Finds the direction the robot is facing

Usually, the IMU data is matched with the data from other sensors (such as wheel encoders, GPS, and so on) to obtain a better estimate of the robot's pose. An IMU sensor is depicted in Figure 1-26.

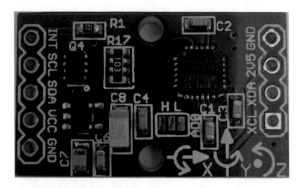

Figure 1-26. *MPU6050 IMU sensor*

Drive Systems for Mobile Robots

The drive system refers to how the wheels are configured to bring about different movements of the robot. There are many ways the wheels of a mobile robot can be arranged and each has strengths and weaknesses.

Drive systems can be broadly classified into holonomic and non-holonomic. A *holonomic* robot can move in X and Y directions in a two-dimensional plane without changing its orientation. On the contrary, *non-holonomic* robots can only move along one axis without changing their orientation. If they need to move along another axis, they first need to change orientation by turning. A holonomic system is used when the robot requires flexibility to move in any direction, whereas non-holonomic robots are used where the movements are mostly along straight lines and curves. See Figures 1-27 and 1-28.

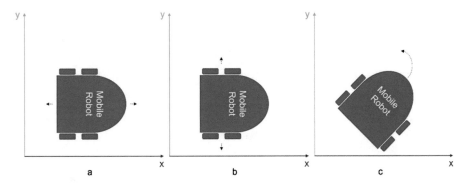

Figure 1-27. *Holonomic robot movements. A) movement (front and back) along the x-axis, b) movement (sideways) along the y-axis (without changing orientation), c) rotation along the z-axis (changing orientation)*

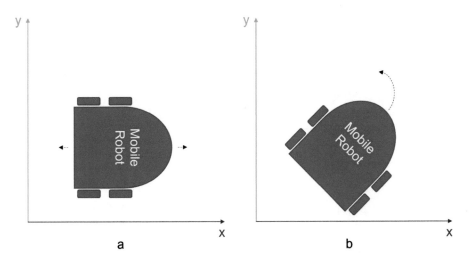

Figure 1-28. *Non-holonomic robot movements. a) movement (front and back) along the x-axis, b) rotation along the z-axis (changing orientation)*

Some of the more popular drive systems are discussed in the following sections.

Differential Drive

Differential drive systems are non-holonomic and consist of two independently controlled wheels capable of rotating at different speeds. There is also one or more free wheels or caster wheels to support the structure. This drive system enables the robot to move forward, backward, and rotate, as shown in Figure 1-29.

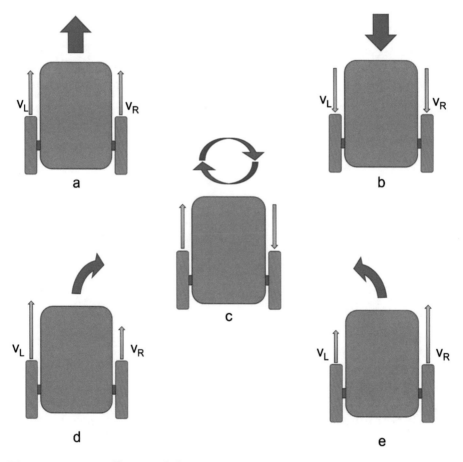

Figure 1-29. *Differential drive system*

Skid Steer

Skid steer drive systems are non-holonomic and are similar to differential drive systems. In contrast to differential drive systems, the caster wheels are replaced with driving wheels. A skid steer drive system often has tracks on either side or four (or more) independently driven wheels. This brings about forward, backward, and rotational movements. See Figure 1-30.

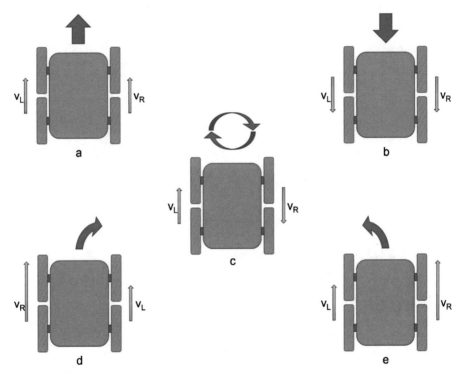

Figure 1-30. *Skid steer drive system*

Tricycle

Tricycle drive systems are non-holonomic and have one front wheel and two rear wheels. The front wheel is steerable, which allows the robot to rotate. The rear wheels are connected using a shaft and are rotated using a single motor. See Figure 1-31.

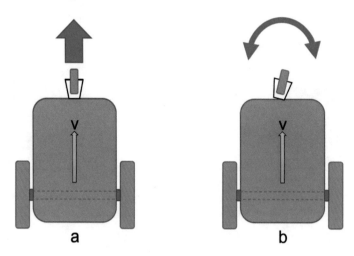

Figure 1-31. *Tricycle drive system*

Ackermann

Ackermann drive systems are non-holonomic and have four wheels. The wheels on the front are steerable, thus allowing the robot to change orientation. The rear wheels are the driving wheels and are joined by a shaft. They are controlled by a single motor. See Figure 1-32.

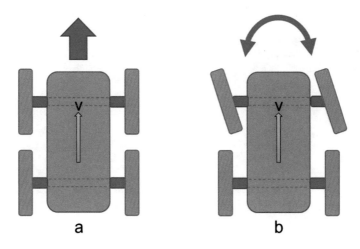

Figure 1-32. *Ackermann drive system*

Omni-Directional

Omni-directional drive systems are holonomic and these robots use a special set of wheels called omni wheels. An *omni wheel* has several rollers along the circumference, perpendicular to the wheel's axis of rotation. These wheels allow the robot to move in any direction without changing orientation. There are three-wheel and four-wheel omni-directional drive configurations. In a three-wheel configuration, three omni wheels are arranged 120 degrees apart on a robot base. Whereas in the four-wheel configuration, four omni wheels are placed 90 degrees apart from each other on a robot base. Each wheel is independently drivable. See Figure 1-33.

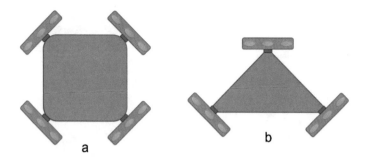

Figure 1-33. *Omni directional drive system a) four-wheel configuration, b) three-wheel configuration*

Mecanum

Mecanum drive systems are holonomic and use special wheels called *Mecanum wheels,* which include a main wheel with small rollers arranged along the circumference. These small rollers are placed at 45 degrees to the axis of rotation of the main wheel. Also, each wheel is independently drivable. This allows the robot to move in any direction without changing orientation. See Figure 1-34.

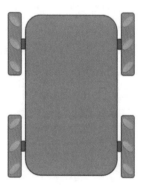

Figure 1-34. *Mecanum wheel drive system*

Ball-bot

Ball-bot drive systems are holonomic and use a single, big spherical wheel attached to the robot base, for driving the robot. The robot self-balances on this wheel, which allows the robot to move in any direction on a surface. See Figure 1-35.

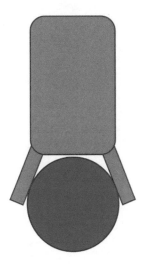

Figure 1-35. *Ball-bot drive system*

Summary

This chapter looked into the history of robotics and the different generations of robots. It also covered the basic mathematics and kinematics of mobile robots, as well as the electronics and drive systems of robots. The next chapter covers the basics of 3D modeling, 3D printing, using Linux, programming, and autonomous robot navigation.

CHAPTER 2

Introduction to Robotics: Part II

Outline

This chapter covers the basics of the following topics:

- 3D modeling

- 3D printing

- Linux

- Programming

- Autonomous robot navigation

Basic 3D Modeling and 3D Printing

3D modeling is the process of creating a three-dimensional representation of an object, while 3D printing is a technique that converts a 3D software model into a real physical model. The following two sections examine these two methods.

© Rajesh Subramanian 2023
R. Subramanian, *Build Autonomous Mobile Robot from Scratch using ROS*,
Maker Innovations Series, https://doi.org/10.1007/978-1-4842-9645-5_2

3D Modeling

You can use 3D modeling to design your robot and use the model in simulation. To model an object, you generally begin with basic shapes and modify them to suit your needs. For example, to design a wheel, you may start with a cylinder shape, remove unwanted areas, smooth the edges, place screw holes, and so on. The example in Figure 2-1 shows different parts of a robot modelled separately and placed together using a 3D modeling software called Tinkercad.

After you model the robot, you can use it for simulation by one of two methods. In the first method, you directly export the complete robot model into a format known as Unified Robot Description Format (URDF). 3D modeling tools such as Fusion 360 have a built-in tool to convert the robot model into URDF format. In the second method, you need to create a basic URDF of the robot, then export the 3D models of the robot parts separately and include them in the URDF file manually.

There is a variety of 3D modeling software available on the market. Some of the popular 3D modeling tools include Tinkercad, Fusion 360, Solidworks, FreeCAD, Blender, and Catia.

Tinkercad is a very easy and beginner-friendly 3D modeling software program (see Figure 2-1). Tinkercad is best for beginners, as it allows them to understand the basics of 3D modeling and quickly design the models. It is available online as a free-to-use web application for 3D modeling. See `www.tinkercad.com/`.

Fusion360 is a professional 3D modeling tool used in the industry. There is a free version of the software available for personal use. For commercial modeling, there are various paid versions. It can be installed on Windows and macOS operating systems. See `www.autodesk.com/products/fusion-360`.

SolidWorks is also another professional designing software program used by engineers in the industry. It offers a free trial version. It can only be installed on the Windows operating system. See `www.solidworks.com/`.

FreeCAD is an open-source 3D modeling software which is free to use. It can be installed on Windows, macOS, and Linux operating systems. See `www.freecadweb.org/`.

Blender is an open-source, free-to-use 3D modeling, animation, and rendering tool. It is mainly used to create animated films, visual effects, and so on. However, it can also be used to create robot models. It can be installed on Windows, macOS, and Linux operating systems. See `www.blender.org/`.

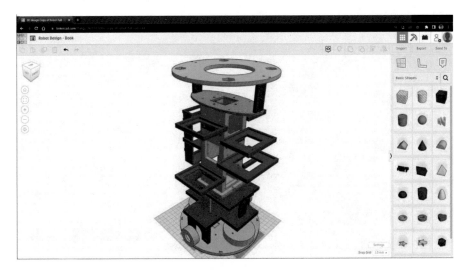

Figure 2-1. *Robot parts individually modelled and placed together using Tinkercad software*

3D Printing

After you design a 3D model, you need to convert the model into a format known as G-code. This process is called *slicing*. The G-code file contains the instructions for the 3D printer to physically generate the designed part. These instructions contain the movements of the printer head, the temperature settings of the hot-end (the part of the printer that melts the filament), printer head speed, and so on. There are several slicing software programs available, of which Ultimaker Cura is a popular choice and can be used to convert 3D model files into G-code. The Cura slicing software is shown in Figure 2-2.

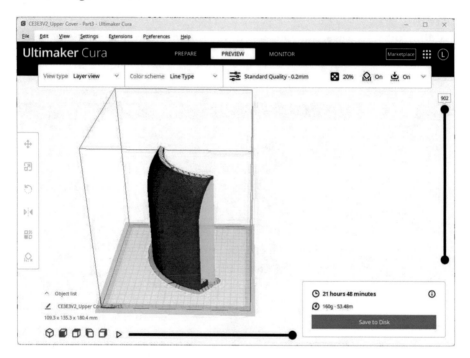

Figure 2-2. *Cura slicing software*

Cura has various parameters that can be adjusted to achieve balance between printing quality and speed. The key adjustable parameters are shown in Figure 2-3. Consider some of the key parameters in more detail:

- **Layer height**: Specifies the height of each printed layer. Lower heights tend to increase the quality of printing, but take more time to print and vice versa.

- **Wall thickness**: Specifies the thickness of the outer layer of the model. Thicker walls make the print stronger.

- **Infill density**: Specifies the amount of material used to fill the inside of the part. For instance, if the part is meant only for aesthetics, we may trade off the strength by minimizing or avoiding infill. Whereas a sturdy part requires a good infill density.

- **Infill pattern**: Specifies how the inside of the model is filled. There are several patterns available that affect the strength and time consumption of the print. For example, line infill does not use much material and is fast to print, but lacks enough strength. Whereas the grid infill pattern provides good strength, but is slower to print and consumes more material.

- **Printing temperature**: Specifies the temperature for printing. It depends on the type of printing material used. For example, the ideal temperature for PLA is around 190° to 220°.

- **Bed temperature**: Specifies the temperature of the bed. This allows the part to stick to the bed while printing happens and also prevents warping.

- **Print speed**: Specifies how fast the printing happens. Faster print speeds tend to reduce the quality of the printing and vice versa.

- **Enable retraction**: Pulls the filament away from the part when printing does not happen. This prevents thin strings of printing material from appearing on the part.

- **Z hop when retracted**: Moves the build plate (bed) down by a specified amount. This prevents the nozzle from hitting the part and avoids strings from occurring on the part.

- **Enable print cooling**: Ensures that the previous layer of melted material is hardened enough so that the nozzle can deposit the next layer. This significantly improves the print quality.

- **Fan speed**: Specifies the speed of the cooling fan. This determines how fast the melted material is hardened.

- **Generate support**: Specifies whether a support structure needs to be printed to print overhanging areas of the model properly. Without support, overhanging areas may collapse.

- **Build plate adhesion type**: This specifies a small layer that is printed on the bed before printing the actual part. This helps to ensure that the actual part sticks firmly to the bed and prevents any warping. There are several adhesion types, such as skirt, brim, raft, and so on.

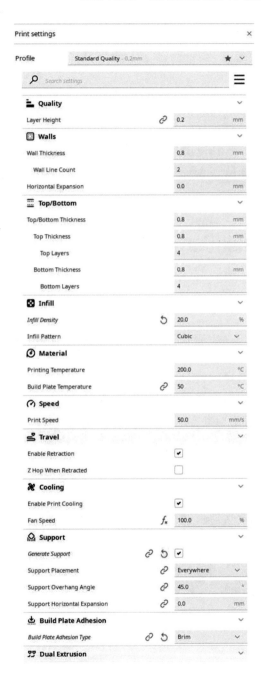

Figure 2-3. *Parameters in the Cura slicing software program*

Basic Linux

Linux is a free, open-source, and very popular operating system used for a wide variety of applications. Linux is widely used in devices such as laptops, smartphones, desktops, tablets, modems, and more. There are several variants of Linux—Ubuntu, Linux Mint, Debian, Fedora, and Manjaro—depending on the application type. For robotics applications using ROS, Ubuntu is widely used. For this reason, this book also uses Ubuntu. The next sections looks at some of the basic Linux commands.

Basic Linux Commands

adduser: Adds a new user

Example:

```
rajesh@ubuntu:~$ sudo adduser swathi
```

The output is shown in Figure 2-4.

```
rajesh@ubuntu:~$ sudo adduser swathi
Adding user `swathi' ...
Adding new group `swathi' (1001) ...
Adding new user `swathi' (1001) with group `swathi' ...
The home directory `/home/swathi' already exists.  Not copying from `/etc/skel'.
New password:
Retype new password:
passwd: password updated successfully
Changing the user information for swathi
Enter the new value, or press ENTER for the default
        Full Name []: Swathi Rajesh
        Room Number []: 1
        Work Phone []: +919656609215
        Home Phone []: NA
        Other []: NA
Is the information correct? [Y/n] y
rajesh@ubuntu:~$
```

Figure 2-4. *Adding a new user*

In Figure 2-4, you can see that, after entering the adduser command, some information is prompted by the operating system. After entering this information, Ubuntu creates a new user.

You can verify whether the OS successfully created the new user by entering the command shown in Figure 2-5.

```
rajesh@ubuntu:~$ cat /etc/passwd | grep swathi
      :x:1001:1001:Swathi Rajesh,1,+919656609215,NA,NA:/home/      :/bin/bash
rajesh@ubuntu:~$ █
```

Figure 2-5. *Verifying the new user*

apt-get: Installs, removes, or upgrades software

Example:

```
rajesh@ubuntu:~$ sudo apt-get install vlc
```

The output is shown in Figure 2-6.

```
Setting up vlc (3.0.9.2-1) ...
Processing triggers for mime-support (3.64ubuntu1) ...
Processing triggers for hicolor-icon-theme (0.17-2) ...
Processing triggers for gnome-menus (3.36.0-1ubuntu1) ...
Processing triggers for libc-bin (2.31-0ubuntu9.9) ...
Processing triggers for man-db (2.9.1-1) ...
Processing triggers for desktop-file-utils (0.24-1ubuntu3) ...
Processing triggers for libvlc-bin:amd64 (3.0.9.2-1) ...
rajesh@ubuntu:~$ █
```

Figure 2-6. *VLC media player installation completed successfully*

cat: Views the contents of a file, creates a new file, concatenates the contents of two files, saves data into another file, displays data on the terminal, and so on.

Example:

```
rajesh@ubuntu:~$ cat file_1 file_2 > combined_file
```

In this example, the contents of two preexisting files—file_1 and file_2—are merged into a new file called combined_file.

The output is shown in Figure 2-7.

```
rajesh@ubuntu:~$ cat file_1    rajesh@ubuntu:~$ cat file_2
1111111111                     2222222222
rajesh@ubuntu:~$               rajesh@ubuntu:~$



                    rajesh@ubuntu: ~ 56x6
rajesh@ubuntu:~$ cat combined_file
1111111111
2222222222
rajesh@ubuntu:~$
```

Figure 2-7. *The contents of each file are displayed using the cat command. The bottom window shows the combined file*

cd: Navigates to the specified directory
Example:

```
rajesh@ubuntu:~$ cd ROS_Book_WS/
```

The output is shown in Figure 2-8.

```
rajesh@ubuntu:~$ cd ROS_Book_WS/
rajesh@ubuntu:~/ROS_Book_WS$
```

Figure 2-8. *Changed the current directory to the specified one*

chmod: Changes the permissions of a file such as read/write/execute.
Example:

```
rajesh@ubuntu:~$ sudo chmod +x robot.py
```

The output is shown in Figure 2-9.

```
⊞               rajesh@ubuntu: ~ 51x3
rajesh@ubuntu:~$ ls -l robot.py
-rw-rw-rw- 1 rajesh rajesh 36 Jan 18 21:07 robot.py
rajesh@ubuntu:~$ ▯
```

```
⊞               rajesh@ubuntu: ~ 51x3
rajesh@ubuntu:~$ sudo chmod +x robot.py
[sudo] password for rajesh:
rajesh@ubuntu:~$ ▯
```

```
▪               rajesh@ubuntu: ~ 51x4
rajesh@ubuntu:~$ ls -l robot.py
-rwxrwxrwx 1 rajesh rajesh 36 Jan 18 21:07 robot.py
rajesh@ubuntu:~$
```

Figure 2-9. *From top: Displaying the current attributes of the file, changing the execute permission of the file, executing permissions given to the file*

Here, you can see the letters rwx after executing the sudo chmod +x robot.py command. They stand for read, write, and execute permissions, respectively. The file has been granted the execute permission.

chown: Changes the ownership of files or folders to the specified user.

An example is shown in Figure 2-10.

```
rajesh@ubuntu:~$ ls -l | grep file_1
-rw-rw-r-- 1 rajesh rajesh   11 Jan 18 08:20 file 1
rajesh@ubuntu:~$ sudo chown swathi file_1
```

Figure 2-10. *Changing the ownership of file_1*

In Figure 2-10, you can see that after executing the ls -l | grep file_1 command, the current owner of the file is rajesh. Then, in the second command, we are changing the ownership of the file to another user called swathi.

55

The output is shown in Figure 2-11.

```
rajesh@ubuntu:~$ ls -l | grep file_1
-rw-rw-r-- 1 swathi rajesh    11 Jan 18 08:20 file_1
```

Figure 2-11. *Verifying that the ownership has been changed*

In Figure 2-11, you can see that the ownership of the file has been correctly changed to swathi.

clear: Clears the contents of the terminal window. You can also use the shortcut Ctrl+l. An example is shown in Figure 2-12.

```
rajesh@ubuntu:~$ abc

Command 'abc' not found, did you mean:

  command 'abx' from deb abx (0.0~b1-1build1)
  command 'ajc' from deb aspectj (1.9.2-1)
  command 'ab' from deb apache2-utils (2.4.41-4ubuntu3.12)
  command 'ac' from deb acct (6.6.4-2)
  command 'abe' from deb abe (1.1+dfsg-3)
  command 'arc' from deb arc (5.21q-6)
  command 'arc' from deb arcanist (0~git20190207-1)
  command 'nbc' from deb nbc (1.2.1.r4+dfsg-9)
  command 'bc' from deb bc (1.07.1-2build1)
  command 'atc' from deb bsdgames (2.17-28build1)
  command 'aec' from deb libaec-tools (1.0.4-1)
  command 'asc' from deb asc (2.6.1.0-6build4)
  command 'cbc' from deb coinor-cbc (2.10.3+repack1-1build1)
  command 'axc' from deb afnix (2.9.2-2build1)

Try: sudo apt install <deb name>

rajesh@ubuntu:~$ clear
```

Figure 2-12. *Clearing the terminal screen*

The output is shown in Figure 2-13.

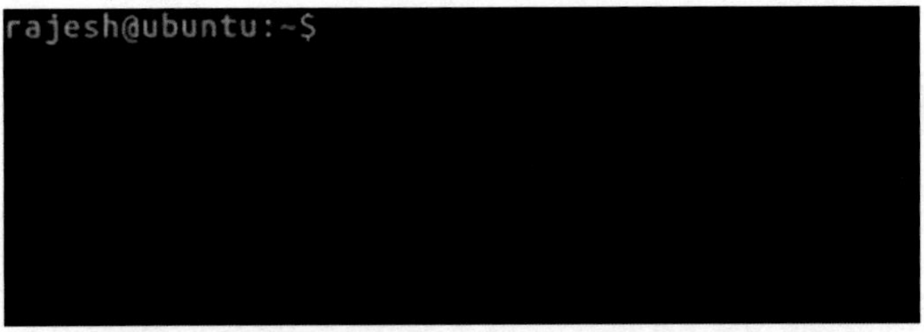

Figure 2-13. *Terminal window has been cleared*

cp: Copies a file or folder to a destination. An example is shown in Figure 2-14.

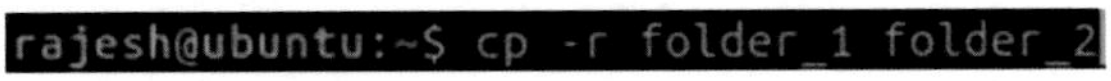

```
rajesh@ubuntu:~$ cp -r folder_1 folder_2
```

The output is shown in Figure 2-15.

```
rajesh@ubuntu:~$ cd folder_2
rajesh@ubuntu:~/folder_2$ ls
file_2   folder_1
rajesh@ubuntu:~/folder_2$
```

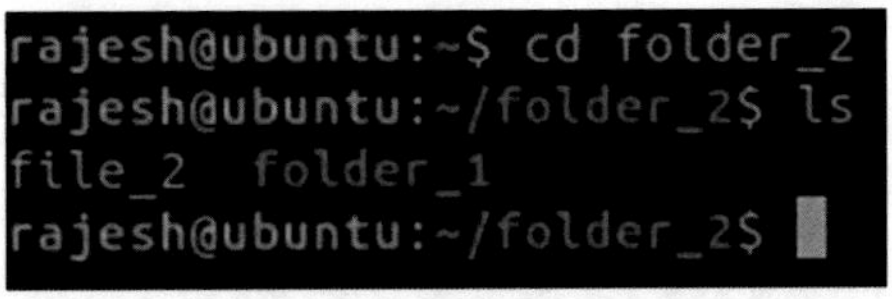

Figure 2-14. *folder_1 copied into folder_2*

diff: Finds the differences between two files or folders. Example:

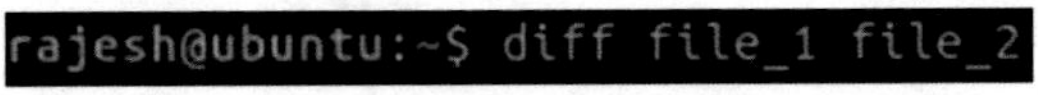

```
rajesh@ubuntu:~$ diff file_1 file_2
```

The output is shown in Figure 2-15.

```
rajesh@ubuntu:~$ diff file_1 file_2
1c1
< 1111111111
---
> 2222222222
rajesh@ubuntu:~$
```

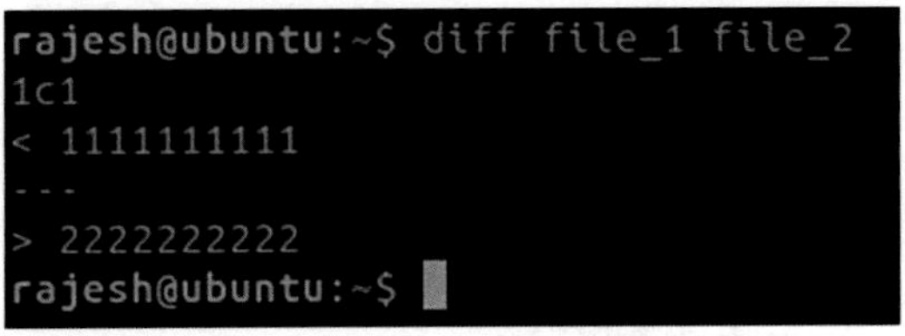

Figure 2-15. *Differences between the two files*

echo: Displays text on the terminal or writes to a file. For example:

```
rajesh@ubuntu:~$ echo "This is a book on building robots" > file_1
```

The output is shown in Figure 2-16.

```
rajesh@ubuntu:~$ echo "This is a book on building robots" > file_1
rajesh@ubuntu:~$ cat file_1
"This is a book on building robots"
rajesh@ubuntu:~$
```

Figure 2-16. *String written to file*

exit: Exits from the current terminal window. For example:

```
rajesh@ubuntu:~$ exit
```

The output is that the current terminal window is closed.

grep: Searches for a specified pattern of characters in a file and prints the lines containing the pattern. For example:

```
rajesh@ubuntu:~$ grep "robot" file_1
```

The output is shown in Figure 2-17.

```
rajesh@ubuntu:~$ grep "robot" file_1
"This is a book on building robots"
rajesh@ubuntu:~$
```

Figure 2-17. *The string is found and displayed*

head: Displays the first specified "n" characters of a file. For example:

```
rajesh@ubuntu:~$ head -n 5 file_1
```

The output is shown in Figure 2-18.

```
rajesh@ubuntu:~$ cat file_1
File Begins
1
2
3
4
5
6
7
8
9
10
File Ends
rajesh@ubuntu:~$ head -n 5 file_1
File Begins
1
2
3
4
rajesh@ubuntu:~$
```

Figure 2-18. *The first five lines of the file*

history: Displays the previously executed commands in the terminal. For example:

```
rajesh@ubuntu:~$ history
```

The output is shown in Figure 2-19.

```
1002   ls
1003   cp -r folder_1 folder_2
1004   cd folder_2
1005   ls
1006   cd ..
1007   ls
1008   diff file_1 file_2
1009   echo "This is a book on building robots" > file_1
1010   cat file_1
1011   ls
1012   echo "This is a book on building robots" > file_1
1013   cat file_1
1014   grep "robot" file_1
1015   cat file_1
1016   grep "robot" file_1
1017   grep "robots" file_1
1018   grep "robot" file_1
1019   head file_1
1020   cat file_1
1021   head -n 5 file_1
1022   cat file_1
1023   head -n 5 file_1
1024   history
rajesh@ubuntu:~$
```

Figure 2-19. *History of previously execu ted commands*

kill: Terminates a running process. This is typically used when a program is unresponsive. An example is shown in Figure 2-20.

```
rajesh@ubuntu:~$ cat python_script.py
while True:
    print("infinite loop")
rajesh@ubuntu:~$ python3 python_script.py
```

Figure 2-20. *An infinite loop*

When the program shown in Figure 2-20 runs, it causes an infinite loop. To terminate it, you can enter the command shown in Figure 2-22. The process ID can be obtained by using the process monitor, as shown in Figure 2-21.

Process Name ▼	User	% CPU	ID	Memory	Disk read tot;	Disk wri
ibus-dconf	rajesh	0	1705	636.0 KiB	12.0 KiB	
ibus-engine-simple	rajesh	0	1871	632.0 KiB	N/A	
ibus-extension-gtk3	rajesh	0	1706	10.8 MiB	744.0 KiB	
ibus-portal	rajesh	0	1712	652.0 KiB	N/A	
ibus-x11	rajesh	0	1710	5.9 MiB	N/A	
nautilus	rajesh	0	3765	21.1 MiB	N/A	236.
pulseaudio	rajesh	0	1462	4.6 MiB	272.0 KiB	8.
python3	rajesh	35	4514	3.3 MiB	N/A	
(sd-pam)	rajesh	0	1457	3.5 MiB	N/A	
snap-store	rajesh	0	1940	124.3 MiB	52.0 MiB	9.2
ssh-agent	rajesh	0	1632	452.0 KiB	N/A	
systemd	rajesh	0	1456	2.3 MiB	29.3 MiB	5.5
terminator	rajesh	26	3906	24.8 MiB	N/A	69.9
tracker-miner-fs	rajesh	0	1465	8.0 MiB	320.0 KiB	
update-notifier	rajesh	0	2910	6.4 MiB	4.6 MiB	6.9
vmtoolsd	rajesh	0	1853	11.2 MiB	9.6 MiB	
xdg-desktop-portal	rajesh	0	2062	1.2 MiB	804.0 KiB	
xdg-desktop-portal-gtk	rajesh	0	2066	5.9 MiB	592.0 KiB	
xdg-document-portal	rajesh	0	1962	612.0 KiB	220.0 KiB	

Figure 2-21. *Getting the process ID*

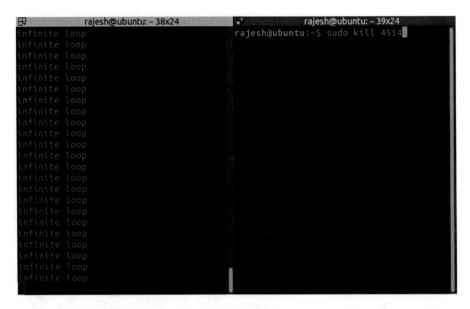

Figure 2-22. *Killing the unresponsive process*

The output is as follows:

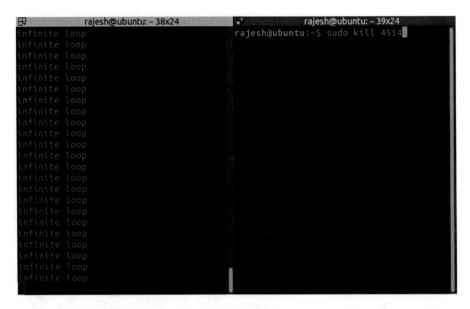

locate: Finds the location of a specified file. For example:

```
rajesh@ubuntu:~$ locate file_1
```

The output is shown in Figure 2-23.

```
rajesh@ubuntu:~$ locate file_1
/home/rajesh/file_1
/home/rajesh/folder_1/file_1
rajesh@ubuntu:~$
```

Figure 2-23. *Files found with a specified name*

ls: Displays all the contents within a folder. For example:

```
rajesh@ubuntu:~/ROS_Book_WS/src/chapter3/scripts$ ls
```

The output is shown in Figure 2-24.

```
rajesh@ubuntu:~/ROS_Book_WS/src/chapter3/scripts$ ls
custom_msg_publisher.py    move_robot_client.py    subscriber.py
custom_msg_subscriber.py   move_robot_server.py    trigger_switch_client.py
custom_srv_client.py       publisher.py            trigger_switch_server.py
custom_srv_server.py       sample_script.py
rajesh@ubuntu:~/ROS_Book_WS/src/chapter3/scripts$
```

Figure 2-24. *Displaying the contents of the folder*

man: Shows a reference manual of all the commands in Linux. For example:

```
rajesh@ubuntu:~$ man python3
```

The output is shown in Figure 2-25.

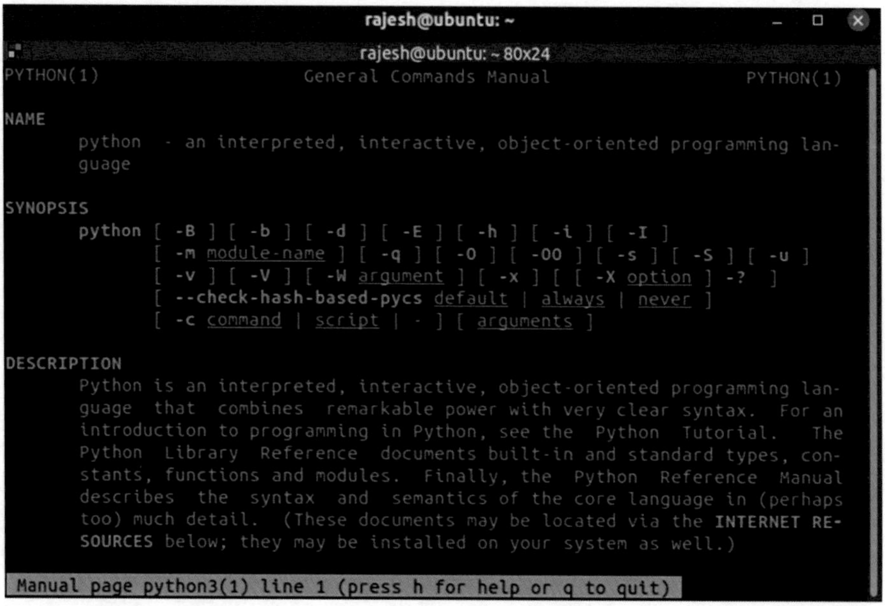

```
                              rajesh@ubuntu: ~                    _  □  ⊗
                         rajesh@ubuntu: ~ 80x24
PYTHON(1)                  General Commands Manual                   PYTHON(1)

NAME
       python  - an interpreted, interactive, object-oriented programming lan-
       guage

SYNOPSIS
       python [ -B ] [ -b ] [ -d ] [ -E ] [ -h ] [ -i ] [ -I ]
              [ -m module-name ] [ -q ] [ -O ] [ -OO ] [ -s ] [ -S ] [ -u ]
              [ -v ] [ -V ] [ -W argument ] [ -x ] [ [ -X option ] -? ]
              [ --check-hash-based-pycs default | always | never ]
              [ -c command | script | - ] [ arguments ]

DESCRIPTION
       Python is an interpreted, interactive, object-oriented programming lan-
       guage  that  combines  remarkable power with very clear syntax.  For an
       introduction  to programming in Python, see the  Python  Tutorial.  The
       Python  Library  Reference  documents built-in and standard types, con-
       stants, functions and modules.  Finally, the  Python  Reference  Manual
       describes  the  syntax  and  semantics of the core language in (perhaps
       too) much detail.  (These documents may be located via the INTERNET RE-
       SOURCES below; they may be installed on your system as well.)

 Manual page python3(1) line 1 (press h for help or q to quit)
```

Figure 2-25. *Documentation of the python3 command*

mkdir: Creates an empty folder. For example:

```
rajesh@ubuntu:~$ mkdir robotics
```

The output is shown in Figure 2-26.

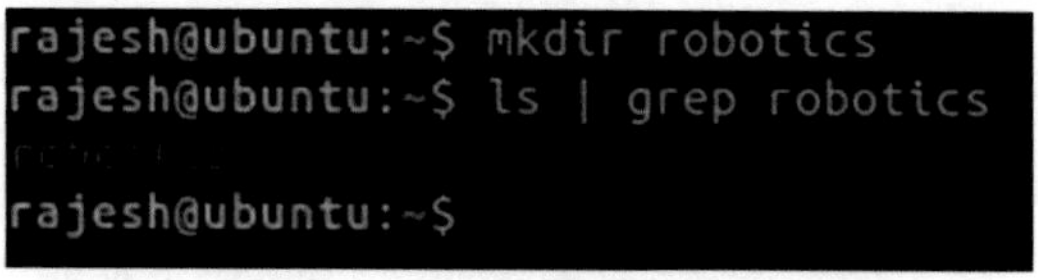

```
rajesh@ubuntu:~$ mkdir robotics
rajesh@ubuntu:~$ ls | grep robotics
rajesh@ubuntu:~$
```

Figure 2-26. *New folder named robotics has been created*

mv: Moves (cuts and pastes) the specified file or folder to the specified location. For example:

```
rajesh@ubuntu:~$ mv file_1 folder_1
```

The output is shown in Figure 2-27.

```
rajesh@ubuntu:~$ cd folder_1
rajesh@ubuntu:~/folder_1$ ls
file_1
rajesh@ubuntu:~/folder_1$
```

Figure 2-27. *File copied to the desired location*

pwd: Displays the path of the current directory. For example:

```
rajesh@ubuntu:~/folder_1$ pwd
```

The output is shown in Figure 2-28.

```
rajesh@ubuntu:~/folder_1$ pwd
/home/rajesh/folder_1
rajesh@ubuntu:~/folder_1$
```

Figure 2-28. *Path of present working directory*

reboot: Restarts the system. For example:

```
rajesh@ubuntu:~$ reboot
```

The output is that the system is rebooted.

rm: Deletes a file or folder. Uses the rm -r command to delete a folder. For example:

```
rajesh@ubuntu:~$ rm file_2
```

The output is as follows:

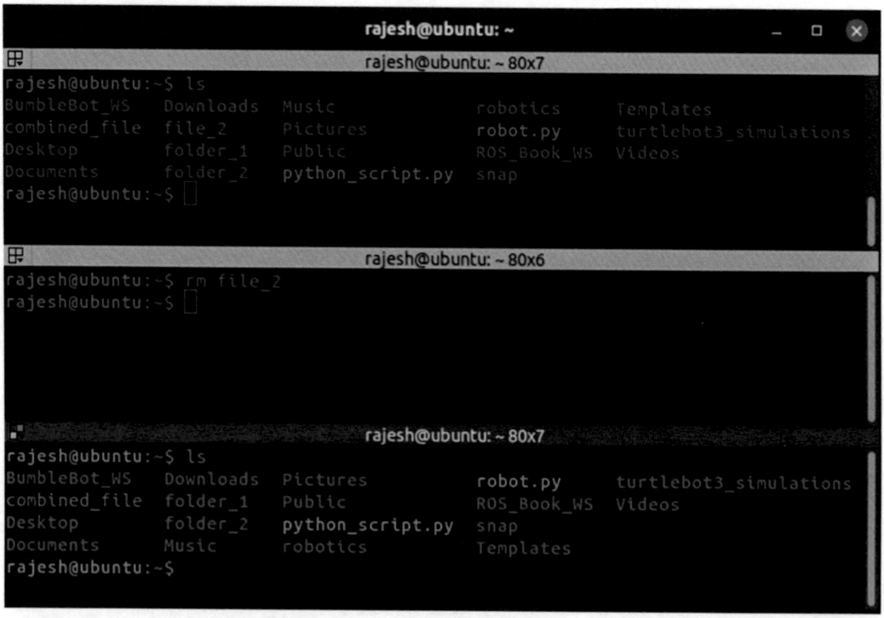

```
                              rajesh@ubuntu: ~                         _   □   ✕

                           rajesh@ubuntu: ~ 80x7
rajesh@ubuntu:~$ ls
BumbleBot_WS     Downloads   Music              robotics      Templates
combined_file    file_2      Pictures           robot.py      turtlebot3_simulations
Desktop          folder_1    Public             ROS_Book_WS   Videos
Documents        folder_2    python_script.py   snap
rajesh@ubuntu:~$


                           rajesh@ubuntu: ~ 80x6
rajesh@ubuntu:~$ rm file_2
rajesh@ubuntu:~$



                           rajesh@ubuntu: ~ 80x7
rajesh@ubuntu:~$ ls
BumbleBot_WS     Downloads   Pictures           robot.py      turtlebot3_simulations
combined_file    folder_1    Public             ROS_Book_WS   Videos
Desktop          folder_2    python_script.py   snap
Documents        Music       robotics           Templates
rajesh@ubuntu:~$
```

rmdir: Deletes an empty folder. For example:

```
rajesh@ubuntu:~$ rmdir folder_1/
```

The output is shown in Figure 2-29.

```
rajesh@ubuntu:~$ rmdir folder_1/
rmdir: failed to remove 'folder_1/': Directory not empty
rajesh@ubuntu:~$
```

Figure 2-29. *Cannot delete the folder, as it is not empty*

shutdown: Shuts down the system. For example:

```
rajesh@ubuntu:~$ shutdown now
```

The output is that the system is shut down.

sudo: Provides elevated rights to perform some operations. For example, when you want to install applications, access system files, run certain commands, and so on. For example:

```
rajesh@ubuntu:~$ sudo apt-get install vlc
```

The output is shown in Figure 2-30.

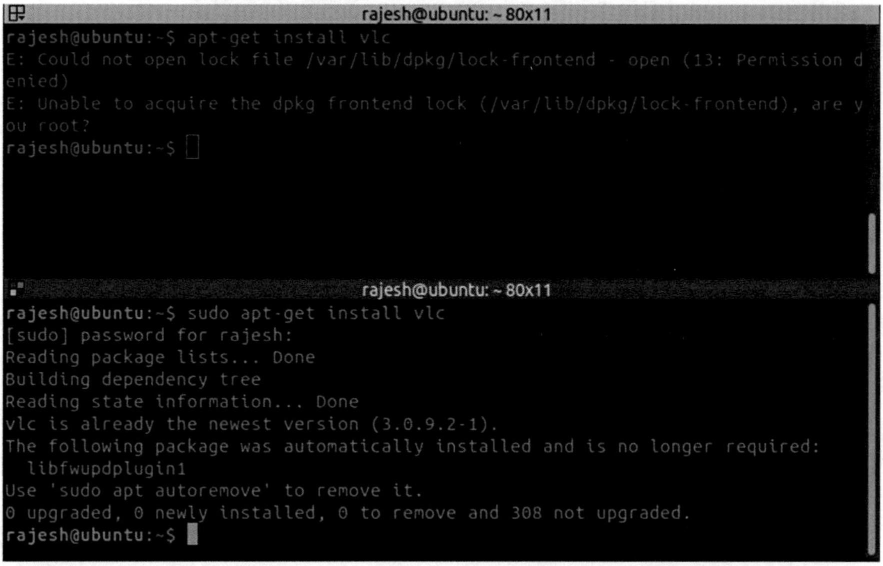

Figure 2-30. *Top window: An error message saying that the command requires elevated permissions. Bottom window: Application installed after entering the sudo command*

tail: Displays the last specified "n" characters of a file. For example:

```
rajesh@ubuntu:~$ tail -n 5 file_1
```

The output is shown in Figure 2-31.

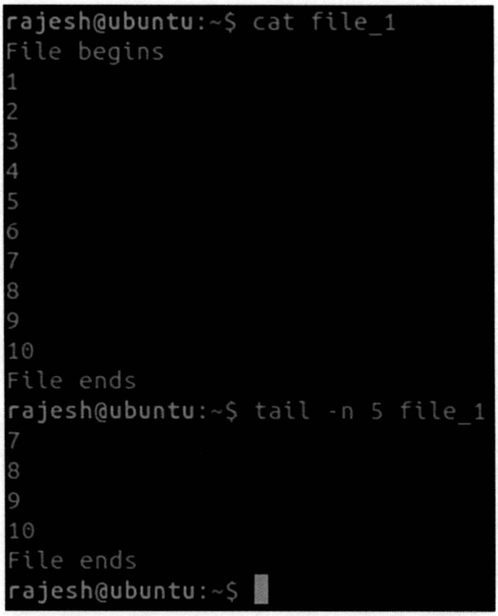

```
rajesh@ubuntu:~$ cat file_1
File begins
1
2
3
4
5
6
7
8
9
10
File ends
rajesh@ubuntu:~$ tail -n 5 file_1
7
8
9
10
File ends
rajesh@ubuntu:~$
```

Figure 2-31. *The last five lines of the file*

touch: Creates an empty file. If a file with the specified name already exists, no operation is performed. For example:

```
rajesh@ubuntu:~$ touch test_file
```

The output is shown in Figure 2-32.

```
rajesh@ubuntu:~$ touch test_file
rajesh@ubuntu:~$ ls
abc              Documents  Music     robotics     Templates                Videos
BumbleBot_WS     Downloads  Pictures  ROS_Book_WS  test_file
Desktop          file_1     Public    snap         turtlebot3_simulations
rajesh@ubuntu:~$
```

Figure 2-32. *Empty file with the specified name*

tree: Displays the contents of the specified folder or the current folder. For example:

```
rajesh@ubuntu:~/ROS_Book_WS/src$ tree
```

The output is shown in Figure 2-33.

```
rajesh@ubuntu: ~/ROS_Book_WS/src 71x27
rajesh@ubuntu:~/ROS_Book_WS/src$ tree
.
├── chapter3
│   ├── action
│   │   └── MoveRobot.action
│   ├── CMakeLists.txt
│   ├── msg
│   │   └── robot_peripherals.msg
│   ├── package.xml
│   ├── scripts
│   │   ├── custom_msg_publisher.py
│   │   ├── custom_msg_subscriber.py
│   │   ├── custom_srv_client.py
│   │   ├── custom_srv_server.py
│   │   ├── move_robot_client.py
│   │   ├── move_robot_server.py
│   │   ├── publisher.py
│   │   ├── sample_script.py
│   │   ├── subscriber.py
│   │   ├── trigger_switch_client.py
│   │   └── trigger_switch_server.py
│   └── srv
│       └── robot_accessories.srv
└── CMakeLists.txt -> /opt/ros/noetic/share/catkin/cmake/toplevel.cmake

5 directories, 17 files
rajesh@ubuntu:~/ROS_Book_WS/src$
```

Figure 2-33. *Contents of the current directory listed in a tree*

userdel: Removes an existing user. For example:

```
rajesh@ubuntu:~$ sudo userdel swathi
```

The output is shown in Figure 2-34.

```
rajesh@ubuntu:~$ cat /etc/passwd | grep swathi
rajesh@ubuntu:~$
```

Figure 2-34. *Verifying the deletion of a user*

You can verify whether the user has been removed successfully by entering the command shown in Figure 2-34.

wc: Displays the number of lines, words, and characters in a file. For example:

```
rajesh@ubuntu:~$ wc file_1
```

The output is shown in Figure 2-35.

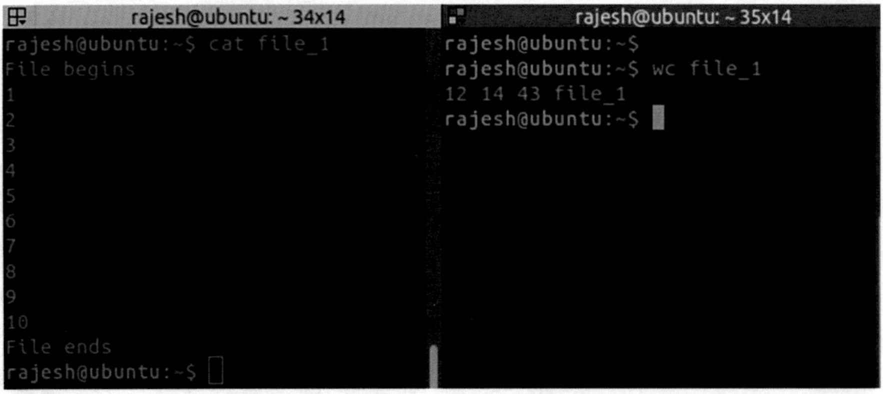

```
rajesh@ubuntu: ~ 34x14                    rajesh@ubuntu: ~ 35x14
rajesh@ubuntu:~$ cat file_1          rajesh@ubuntu:~$
File begins                          rajesh@ubuntu:~$ wc file_1
1                                    12 14 43 file_1
2                                    rajesh@ubuntu:~$
3
4
5
6
7
8
9
10
File ends
rajesh@ubuntu:~$
```

Figure 2-35. *Left: Displaying the contents of the file. Right: Displaying the number of lines, word count, and character count in the file*

Basic Programming

Programming involves writing a set of instructions for carrying out an operation or a set of operations by a computer. In the case of a robot, the operation is carried out by the computer residing inside the robot. This results in the robot performing a task, such as reading sensor inputs, planning a path, moving toward a goal location, and so on. There are several programming languages, including Python, C/C++, Java, and C#. The Robot Operating System (ROS) primarily supports C++ and Python. C++ is a compiled language, that is, the entire code is first translated into machine-readable format before it's executed. Python is an interpreted language, that is, the code is directly executed line by line without converting it into machine code. The main advantage of C++ is the faster speed of execution, whereas Python is comparatively slower. On the other hand, Python codes are very readable, easier to maintain, and can be modified at runtime.

The core concepts of programming include the following.

Variables

Variables are names given to memory locations and are used to store and manipulate data. For example, `a=5,` `b="hello!"`, `c=3.14`, and so on. These values can be changed anytime to suit the needs. For instance, if you want to keep track of the speed of a robot, you can create a variable called `speed`, and then assign the speed value obtained from a sensor.

Data Types

Variables can hold several types of data, such as numbers, letters, decimal numbers, Boolean values, and so on. For example, suppose you want to store goal IDs of the locations your robot would like to traverse in a building. You can represent them using numbers, such as 1 for the kitchen, 2 for the store room, 3 for the main hall, 4 for the dining area, and so on. There are several types of variables:

- **Numeric**: Store numerals including integers and decimal numbers.

- **Character**: Store letters, symbols, and so on.

- **Boolean**: Store true/false values.

- **Array**: Store a list of data of the same type. For example, a list of employee ID numbers, a list of marks obtained by students, and so on.

- **Object**: Contain a collection of mixed data types and operations to be performed. For example, you can create an `elephant` object, which contains data such as `name=jumbo`, `species=asian elephant`, `diet type=herbivore`, and operations such as `eat`, `move`, `find_food`, and so on.

- **Enumerated**: A custom data type and is used to store a set of constant values as a series. For example, if you want to represent the colors of the rainbow as integers from 1 to 7 starting from violet to red respectively, you can create an enumerated data type.

- **Pointer**: Contain memory locations of another variable. They allow you to dynamically access and manipulate the computer's memory, which provides great flexibility.

Operators

You'll often need to perform several operations in your programs (see Table 2-1). The most common operations include the following:

- **Arithmetic**: To perform operations such as addition, subtraction, and so on, using numbers

- **Comparison**: To compare two values and returns a true or false value. For example, equal to (==), greater than (>), lesser than (<), and so on

- **Logical**: To perform logical operations on Boolean variables or expressions. For example, and, or, and not

- **Assignment**: To assign values to a variable in a program. For example, =, +=, -= and so on, where, = assigns a specified value to a variable such as a = 1. The += operator adds the specified value (on the right) to a variable (on the left) and assigns it to the variable (on the left). For example, a += 1. (Note: This is equivalent to writing a = a+1.)

- **Bitwise**: Operates on the individual bits of the specified binary numbers. Some of the bitwise operators are and, or, not, xor, left shift, and so on.

- **Conditional**: Also known as the ternary operator. It evaluates an expression and returns a Boolean value. It is a shorthand representation of an if-else statement.

Table 2-1. *Programming Operators*

Operator	Type	Significance	Example (x=10, y=2)	
			Operation	**Result**
+	Arithmetic	Addition	x + y	12
-		Subtraction	x − y	8
*		Multiplication	x * y	20
/		Division	x / y	5
%		Reminder	x % y	0
^		Power	x ^ y	100
==	Comparison	Equal	x == y	False
!=		Not equal	x != y	True
>		Greater than	x > y	True
<		Less than	x < y	False
>=		Greater than or equal to	x >= y	True
<=		Lesser than or equal to	x <= y	False
&&	Logical	And	(x == 10) && (y == 2)	True
||		Or	(x == 10) || (y == 3)	True
!		Not	! (x == 10)	False
=	Assignment	Assignment	x = 10	10
+=		Addition assignment	x += 5	15
-=		Subtraction assignment	x -= 5	5
*=		Multiplication assignment	x *= 5	50
/=		Division assignment	x /= 5	2
&	Bitwise	And	x & y	2
|		Or	x | y	10

(continued)

Table 2-1. (*continued*)

Operator	Type	Significance	Example (x=10, y=2)	
			Operation	**Result**
~		Not	~x	-11
^		Xor	x ^ y	8
<<		Left shift	x <<	20
>>		Right shift	x >>	5
?:	Conditional	If-else	result = (x > y) ? x : y	10

Branches

Branching is a feature of a program that makes decisions depending on certain conditions or inputs. Branching is commonly done using three methods, discussed next.

If...else

Here, a block of code is executed when the condition is true and another block of code is executed when the condition is false:

```
1. If(condition)
2. {
3.     //Code block 1
4. }
5. Else
6. {
7.     //Code block 2
8. }
```

Else-if Ladder

This is similar to the if-else branching structure, but can handle multiple conditions by executing different blocks of code depending on the condition:

```
1. If(condition1)
2. {
3.      //Code block 1
4. }
5. Else if(condition2)
6. {
7.      //Code block 2
8. }
9. Else if(condition3)
10. {
11.      //Code block 3
12. }
13. .
14. .
15. .
16. Else
17. {
18.      //Code block to be executed when no condition is met
19. }
```

Switch

A switch is very similar to the else-if ladder structure, but can only check a single expression with multiple outcomes:

```
1.  Switch(expression)
2.      Case value1:
3.      {
4.        //Code block 1
5.        Break;
6.      }
7.      Case value2:
8.      {
9.        //Code block 2
10.        Break;
11.      }
12.      Case value3:
13.      {
14.        //Code block 3
15.        Break;
16.      }
17.      .
18.      .
19.      .
20.      Default:
21.      {
22.        //Code block to be executed when no condition is met
23.      }
```

Loops

Loop statements execute a block of code repeatedly for a specified number of times or until a condition is met/broken. There are many types of looping structures, discussed next.

For

This loop allows a block of code to be executed a specified number of times. A for statement has three parts—initialization, a termination condition, and an increment/decrement statement:

```
1. For(i=0; i<n; i++)
2. {
3.    //Code block
4. }
```

Here, i=0 is the initialization, i<n is the termination condition, and 'i++ is the increment part.

While

This structure executes a block of code until a specified condition is met. The loop exits when the condition is broken:

```
1. While(condition)
2. {
3.    //Code block
4. }
```

Do-while

This is very similar to the while loop, but is guaranteed to execute the code block inside the loop at least once, regardless of the condition. The loop breaks when the condition is violated:

```
1. Do
2. {
3.    //Code block
4. } while(condition)
```

Functions

A *function* is a modularized block of code that performs a specific operation. Its modular nature allows it to be reused in various parts of the code. A function typically has a name and zero or more inputs (arguments) and performs some operations. It may also return a value. For example:

```
1. convert_degrees_to_farenheit(temperature)
2. {
3.    farenheit = (temperature × 9/5) + 32
4.    return(farenheit)
5. }
6.
```

Here, `convert_degrees_to_farenheit` is the function name, `temperature` is the input to the function, and the body of the function converts the `temperature` data into Fahrenheit. After conversion, the function returns the Fahrenheit value as the result. Functions help break down the code into smaller, easily manageable units and improve the readability of the code.

Lists

Lists are data structures used to hold a collection of data. That data could be numbers, letters, and objects. For example:

```
colors_list = {"red", "blue", "green", "yellow", "orange"}
```

You can also perform operations on the list, such as adding an element, removing an element, appending an element, sorting elements, and so on.

Dictionaries

Dictionaries are similar to lists, but they include data in the form of key-value pairs rather than individual elements. For example:

```
capital_dict = {"India" : "New Delhi", "Australia" :
"Canberra", "USA" : "Washington DC", ...}
```

The key must be unique in a dictionary. You can perform operations—such as adding, removing, appending, sorting, and so on—on dictionaries.

Files

Files contain of a collection of data stored in a computer disk or any other storage medium. File types can be text, audio, image, video, and so on. The most common types of operations in a file include opening, reading, writing, and closing. By using a suitable programming language, you can perform complex operations such as facial recognition, speech-to-text conversion, and so on, using files.

Classes

Classes contain a collection of mixed data types and operations to be performed. A class is a blueprint of an object and you can create several objects of the class, each with unique data. For example, an `animal` class could contain data such as `name`, `species`, and `diet type`, and operations such as `eat`, `move`, `find_food`, and so on. You could then create an object for a particular animal such as an elephant, a tiger, and so on, with their data. Classes provide a lot of useful features like encapsulation, inheritance, polymorphism, abstraction, overriding, overloading, and so on.

Multi-threading

Multi-threading is a programming technique in which multiple processes are executed simultaneously, instead of sequentially. This improves the overall efficiency of the program. An example of a multi-threaded program is a process waiting to obtain a goal location (from the user) for the robot to navigate to, and another process that plans a path and makes the robot navigate to the specified goal. Multi-threaded programs also have several disadvantages, including increased complexity, deadlocks, race conditions, higher memory utilization, and so on.

Basics of Autonomous Robot Navigation

Autonomous robot navigation is the ability of a robot to move from the initial position to the goal position without colliding with obstacles. Obstacles may be static (walls, partitions, and so on, which are immovable) or dynamic (such as people, chairs, doors and so on, which are subject to changes in position and can move in front of the robot any time in the vicinity of the robot).

There are two types of autonomous navigation in robots:

- Map-based

- Reactive

In *map-based navigation,* the robot uses a previously built map of the environment for navigation. This enables the robot to plan an appropriate path toward the goal. In *reactive navigation,* the robot does not have a map of the environment. Instead, it uses the data obtained from its sensors to gather local information about the environment and plan the movements. The reactive method is used to navigate robots in unknown environments.

Map-Based Navigation

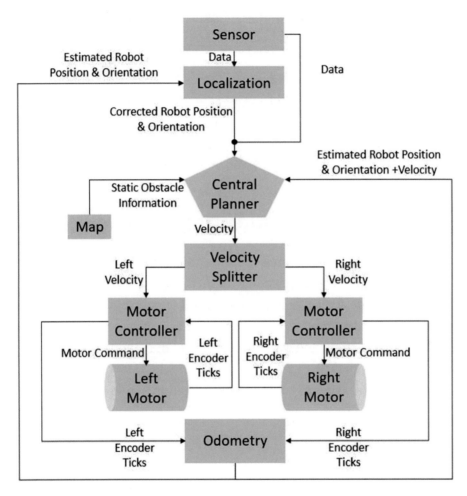

Figure 2-36. *The components of autonomous navigation*

The key components of the map-based navigation method are illustrated in Figure 2-36.

- The sensor node gathers information about non-moveable (static) and moveable (dynamic) obstacles in its vicinity. The obstacle information is passed to the localization node and the central planner.

- The odometry node computes the estimated robot pose by using wheel encoder input (or any other odometry sensors such as accelerometer, gyroscope, GPS, lidar, cameras, and so on). This estimated pose is called *odometry* and is used by the localization node and central planner.

- Localization involves finding the robot on the map. It uses odometry and obstacle information from the sensor to estimate the robot's pose on the map. The localization information is passed to the central planner.

- The map provides information about the static obstacles in the robot's working environment and this is passed to the central planner.

- A central planner has two components—a global path planner and a local path planner. The global planner computes the global path, which connects the starting and goal points. The local planner computes the local paths to closely follow the global path. The central planner takes the map data, localization data, and odometry data as inputs and computes the velocity of the robot. The new velocity is then passed to the velocity splitter.

- The velocity splitter obtains the velocity from the central planner and computes individual wheel velocities. These wheel velocities are passed to the motor controllers.

- Motor controllers take in the velocity values obtained from the velocity splitter and convert them to actual motor commands, which are passed to the motors. The motor controllers also obtain encoder ticks (counts) from the motors and pass them to the odometry node.

- Motors accept the motor commands and rotate the wheels accordingly. Motors also provide encoder pulses (to estimate the wheel positions), which are passed to the motor controller.

Mapping

Mapping involves creating a representation of the robot's working environment. A map enables the robot to perform various actions, such as localization (estimating the pose of the robot on the map), planning the path toward a goal, performing autonomous navigation, and so on. There are a variety of mapping methods in robotics:

- **Occupancy grid mapping**: Involves representing the environment as a grid of several cells (which represent small spaces in the environment). The cells are then marked as occupied or free by using sensor information as the robot explores.

- **SLAM**: Stands for Simultaneous Localization and Mapping. As the name indicates, it is a technique to create a map and at the same time estimate the robot's position within the map. Using the SLAM algorithm, the robot continuously estimates its position and orientation as it traverses and updates the map with the latest obstacle information in the environment gathered using its sensors. The sensors used for mapping typically involve lidar, camera, and odometry sensors (such as wheel encoder, IMU, GPS, and so on).

- **3D Mapping**: A three-dimensional map is generated in contrast to a top-down view obtained from a 2D mapping algorithm. A 3D map will include the height and shape of objects in the robot's environment. Sensors used for 3D mapping involve 3D lidar, stereo camera, and depth camera, along with odometry sensors like IMU, wheel encoders, GPS, and so on.

- **Visual SLAM**: A variant of the SLAM mapping method, this uses the images obtained from a camera to generate the map. Typical cameras used for visual SLAM are monocular cameras, stereo cameras, and depth cameras. Odometry sensors can also be used along with cameras for better accuracy.

- **Multi-Sensor Fusion**: In this method, the data from different sensor sources are combined to create a map that has better accuracy. The sensors used can include lidar, camera, sonar, and so on.

Navigation Algorithms

Navigation algorithms are collections of algorithms that work together to bring about autonomous navigation in robots. They include:

- **Mapping**: Create a representation of the robot's working environment. A map contains the static obstacles and the overall outline of the environment. A map can be 2D or 3D depending on the sensors and algorithms used for mapping. Various features on the map may also be represented by different colors. Some of the algorithms used for mapping include:

  - Gmapping

  - HectorSLAM

  - KartoSLAM

  - RTAB-Map

  - OctoMap

- **Localization**: To estimate the position and orientation of the robot in the map. Some popular algorithms used for localization include:

  - Adaptive Monte Carlo Localization (AMCL)

  - Extended Kalman Filter (EKF)

- **Path planning**: Determine a path from the current pose to the target pose, avoiding the static obstacles. Path planner typically uses a map of the environment to generate the best path for the robot. The path generated by the planner is a set of waypoints (i.e., intermediate goal points). Various algorithms used for path planning include:

  - A*

  - Dijkstra's

  - Rapidly exploring Random Tree (RRT)

- Probabilistic Road Map (PRM)

- RRT*

- Generalized Voronoi Diagrams (GVD)

- Gradient Descent Algorithm (GDA)

- **Motion planning**: Generate intermediate paths to follow the waypoints generated by the path planner. This also takes into account any robot constraints (such as robot geometry, maximum velocity/acceleration, joint constraints, and so on), obstacle position, free spaces, inflation around obstacles (to provide a safety radius around obstacles to avoid collision), and so on. This allows robots to generate a smooth and safe path to follow, to reach the goal. Some motion planners include:

  - Base local planner

  - Trajectory planner

  - Teb local planner

  - Dynamic-Window Approach (DWA)

  - Dynamic-Window Based (DWB)

- **Control**: Generate actuator commands to physically move the robot according to the plan generated by the motion planner. Some control algorithms include:

  - Diff drive controller

  - Skid steer controller (a variant of diff drive controller)

  - Mecanum controller (a variant of diff drive controller)

- Ackermann steering controller

- Four-wheel steering controller

- **Obstacle avoidance**: Avoid any dynamic obstacles appearing while the robot traverses toward the goal. Some dynamic obstacle avoidance algorithms include:

  - Potential field methods

  - Receding horizon control

  - Model predictive control

Summary

This chapter looked at the basics of 3D modeling, 3D printing, Linux, programming, and autonomous robot navigation. The next chapter is about setting up a workstation for simulation by installing Ubuntu, ROS, VS Code, and other software.

CHAPTER 3

Setting Up a Workstation for Simulation

Outline

This chapter shows you how to set up a workstation for simulation and install some useful tools, including the following:

- Ubuntu

- ROS Noetic

- Visual Studio Code

Workstation Setup

Often, a robot uses a less powerful computer like a Raspberry Pi for its operations, which is not suitable for simulation. Moreover, a graphics card is required for the simulation to work smoothly for rendering, image simulation, and so on. A workstation is a separate computer that is usually more powerful and has a dedicated graphics card for simulating the robot.

© Rajesh Subramanian 2023
R. Subramanian, *Build Autonomous Mobile Robot from Scratch using ROS*,
Maker Innovations Series, https://doi.org/10.1007/978-1-4842-9645-5_3

Once the robot simulation works correctly in the workstation, you need to port the functionalities into the actual robot. For that, you need to install Ubuntu and ROS, copy the user-defined programs, and set up their dependencies (e.g., Python libraries, additional ROS components, and so on) in the robot. The procedure for installing these software components is similar to the one you follow when setting up the workstation.

Simulation is an important step because, before you build your physical robot, simulation helps you understand the various aspects of the robot. For instance, whether the robot design has any faults, whether the robot moves properly when teleoperated, whether the algorithms work properly, whether the navigation stack is tuned well, and so on.

To simulate and control the robot, you need the following:

- **Operating system**: Helps the robot communicate with its sensors, motor controllers, and other peripherals. Usually, the Ubuntu Linux distribution is employed because it is simple, lightweight, and easy to use.

- **Robot Operating System (ROS)**: ROS is not exactly an operating system, but is a middleware that sits between the actual operating system and various programs that run in the robot. ROS also enables communication between different programs in the robot. ROS is fully compatible with Ubuntu and experimentally supported in other operating systems, such as macOS, Microsoft Windows, and Fedora Linux.

- **IDE** (*optional*): You can install an integrated development environment (IDE) to help reduce errors and improve the quality of the programs you write to control the robot.

Each version of ROS is only fully compatible with a specific version of the Ubuntu operating system. Some popular ROS distributions and their compatible Ubuntu versions are listed in Table 3-1.

Table 3-1. *ROS Distributions and Compatible Ubuntu Versions*

ROS Version	Compatible Ubuntu Version
Indigo Igloo	14.04.06 LTS (Trusty Tahr)
Kinetic Kame	16.04.7 LTS (Xenial Xerus)
Melodic Morenia	18.04.5 LTS (Bionic Beaver)
Noetic Ninjemys	20.04.1 LTS (Focal Fossa)

The next section looks into the procedure for installing the required software components in a workstation. This book uses Ubuntu MATE 20.04 (Focal Fossa) as the operating system, ROS (Noetic Ninjemys) as the ROS version, and Visual Studio Code as the IDE.

Ubuntu MATE (Focal Fossa) Installation

There are three popular ways to install Ubuntu:

1. **Virtual Machine (VM)**: Using this method, Ubuntu can run inside a virtual machine on top of an existing operating system, such as Windows or Mac. It is convenient to set up and configure Ubuntu.

2. **Windows Store App**: Using this method, Ubuntu can be installed in Windows as an application from the Microsoft Store app. Several configurations need to be done and the setup is a bit lengthy. Moreover, most of the interactions with Ubuntu need to be carried

out by typing commands, because it is not possible to install a desktop environment. However, you can run graphical applications using WSLg (Windows Subsystem for Linux GUI). To learn more, visit:

```
https://ubuntu.com/tutorials/install-ubuntu-
on-wsl2-on-windows-10#1-overview
```

3. **Dual-boot**: Using this method, you can install Ubuntu alongside Windows. This process allows the users to choose the operating system during startup. The capability of the system is fully utilized during the simulation and is the ideal method. To learn more, visit:

```
https://help.ubuntu.com/community/
WindowsDualBoot
```

This book installs Ubuntu via a virtual machine. It uses a virtual machine software program called VirtualBox.

Prerequisites

- Download and install VirtualBox from this link:

  ```
  www.virtualbox.org/wiki/Downloads
  ```

- Download the Ubuntu MATE image from this link:

  ```
  https://ubuntu-mate.org/download/amd64/focal/
  ```

Installation Steps

1. Open Oracle VM VirtualBox Manager and click New. See Figure 3-1.

Figure 3-1. *Creating and configuring the virtual disk*

2. Give the virtual machine a name and select Linux for the Type and Ubuntu for the Version (32/64-bit depending on whether you have a 32/64-bit CPU). Finally, select the path of the Ubuntu MATE image you downloaded previously and click Next. See Figure 3-2.

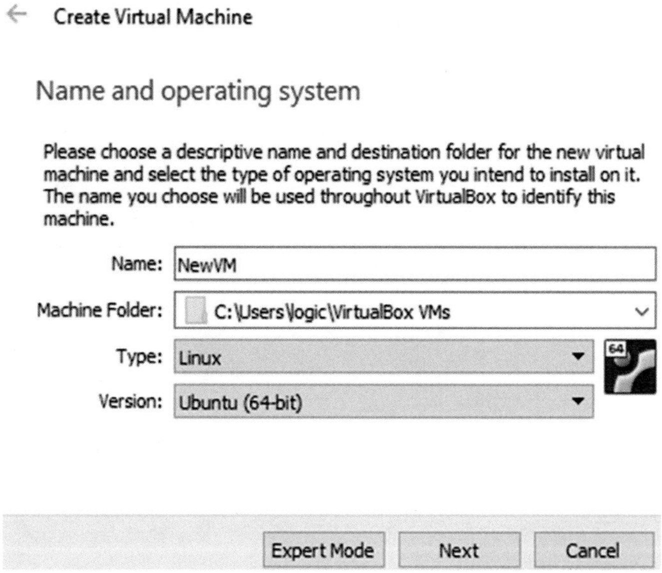

Figure 3-2. *Creating and configuring the virtual disk*

3. Specify the amount of RAM to be used in the virtual machine. You can provide any value of RAM, but a minimum RAM size of 1GB is recommended. The virtual machine works smoother if more RAM is allocated, but affects the performance of the host machine. This example allots 4GB of RAM to the virtual machine. Visit `https://ubuntu-mate.org/about/requirements/`. See Figure 3-3.

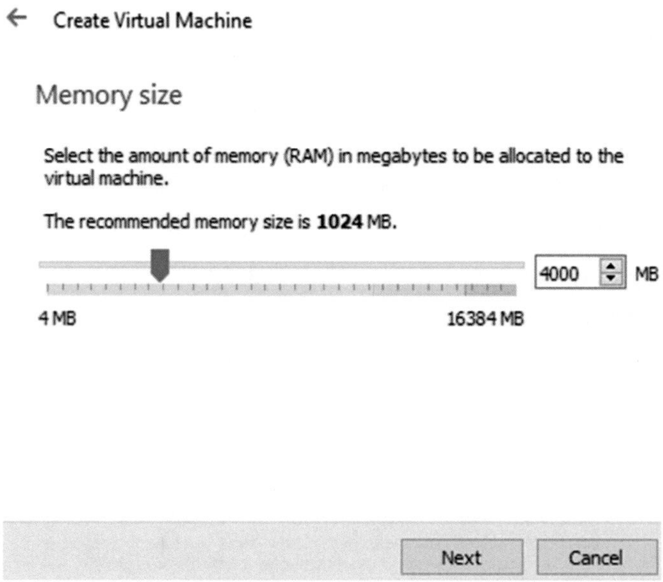

Figure 3-3. *Creating and configuring the virtual disk*

4. Select Create a Virtual Hard Disk Now and then click Create. See Figure 3-4.

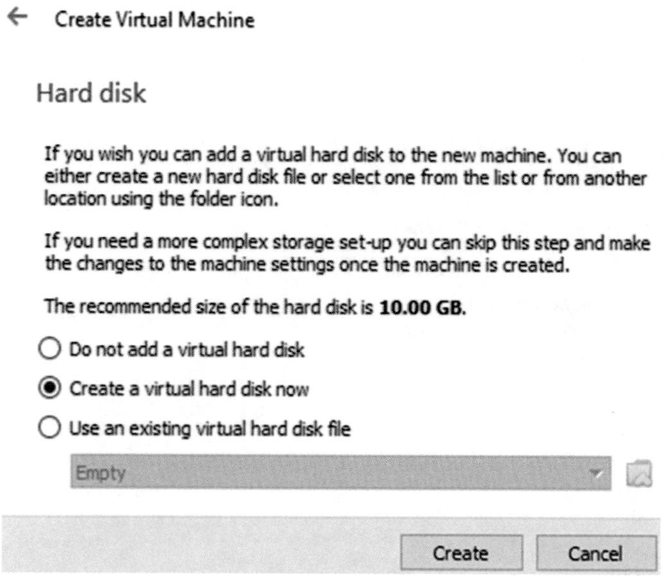

← Create Virtual Machine

Hard disk

If you wish you can add a virtual hard disk to the new machine. You can either create a new hard disk file or select one from the list or from another location using the folder icon.

If you need a more complex storage set-up you can skip this step and make the changes to the machine settings once the machine is created.

The recommended size of the hard disk is **10.00 GB**.

○ Do not add a virtual hard disk

◉ Create a virtual hard disk now

○ Use an existing virtual hard disk file

Empty

Create   Cancel

Figure 3-4. *Creating and configuring the virtual disk*

5. Select VDI (VirtualBox Disk Image) from the Hard disk file type and click Next. See Figure 3-5.

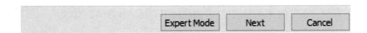

← Create Virtual Hard Disk

Hard disk file type

Please choose the type of file that you would like to use for the new virtual hard disk. If you do not need to use it with other virtualization software you can leave this setting unchanged.

◉ VDI (VirtualBox Disk Image)

○ VHD (Virtual Hard Disk)

○ VMDK (Virtual Machine Disk)

Expert Mode   Next   Cancel

Figure 3-5. *Creating and configuring the virtual disk*

6. Select Fixed Size so that the virtual disk will be faster and then click Next. See Figure 3-6.

← Create Virtual Hard Disk

Storage on physical hard disk

Please choose whether the new virtual hard disk file should grow as it is used (dynamically allocated) or if it should be created at its maximum size (fixed size).

A **dynamically allocated** hard disk file will only use space on your physical hard disk as it fills up (up to a maximum **fixed size**), although it will not shrink again automatically when space on it is freed.

A **fixed size** hard disk file may take longer to create on some systems but is often faster to use.

○ Dynamically allocated

◉ Fixed size

Next   Cancel

Figure 3-6. *Creating and configuring the virtual disk*

Specify the file location and the amount of storage to be allocated for the virtual disk; then click Create. Ubuntu MATE 20.04 requires a minimum 8GB of space and ROS noetic requires 2.7GB of space. Also, you'll need to install additional software and packages. Therefore, consider these factors when deciding on the storage space. This example allocates 30GB of storage space. Visit https:// ubuntu-mate.org/about/requirements/. See Figure 3-7.

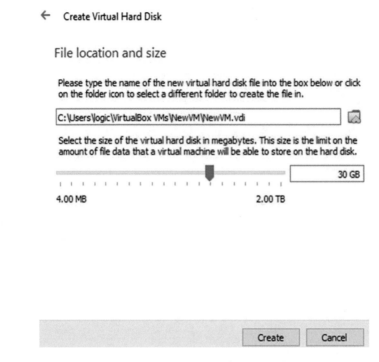

Figure 3-7. Creating and configuring the virtual disk

7. Start the virtual machine by clicking the Start
    button. See Figure 3-8.

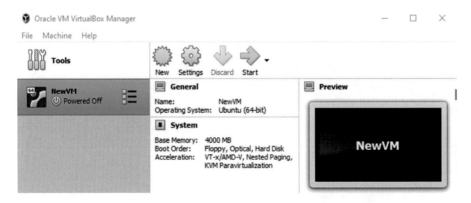

Figure 3-8. Starting the virtual disk

8. Click the Choose a Virtual Optical Disk File button (the yellow folder button). See Figure 3-9.

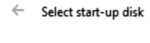

Figure 3-9. *Starting the virtual disk*

9. Click the Add button. See Figure 3-10.

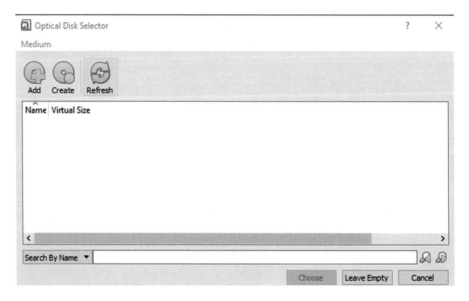

Figure 3-10. *Starting the virtual disk*

10. Select the Ubuntu MATE (Focal Fossa) ISO file you downloaded earlier and click the Open button. See Figure 3-11.

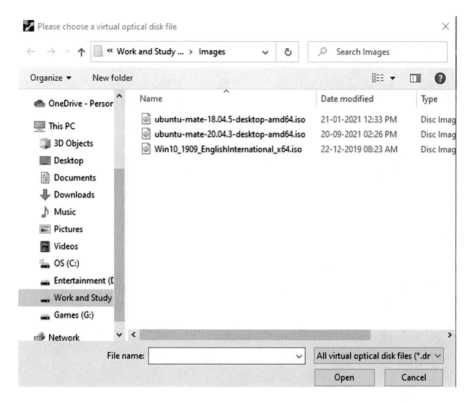

Figure 3-11. *Starting the virtual disk*

11.   Select the ISO file and click Choose. See Figure 3-12.

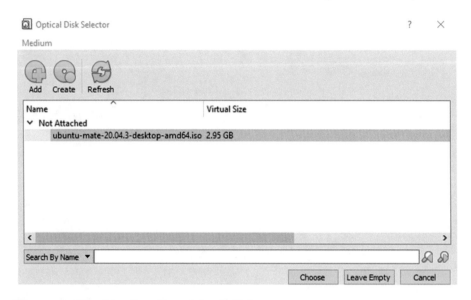

Figure 3-12. *Starting the virtual disk*

12. Click the Start button. See Figure 3-13.

Figure 3-13. *Starting the virtual disk*

13. Click the Install Ubuntu MATE button. See Figure 3-14.

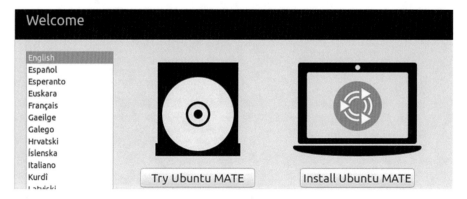

Figure 3-14. *Installing Ubuntu in VirtualBox*

14. Select the appropriate language and then click the
     Continue button. See Figure 3-15.

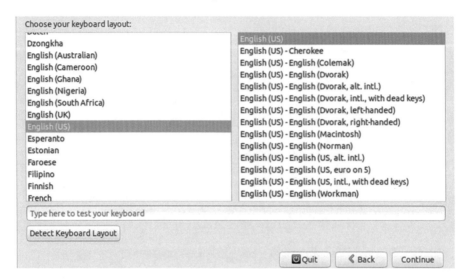

Figure 3-15. *Installing Ubuntu in VirtualBox*

15. Select Normal Installation. Optionally, you can
     check the Download Updates While Installing
     Ubuntu MATE and Install Third-Party Software for
     Graphics and Wi-Fi and Additional Media Formats
     check boxes. See Figure 3-16. Then click Continue.

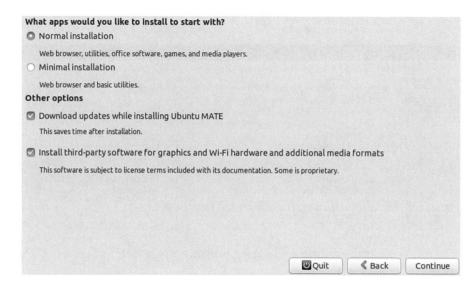

Figure 3-16. *Installing Ubuntu in VirtualBox*

16.   Select Erase Disk and Install Ubuntu MATE; then
       click Install Now. See Figure 3-17.

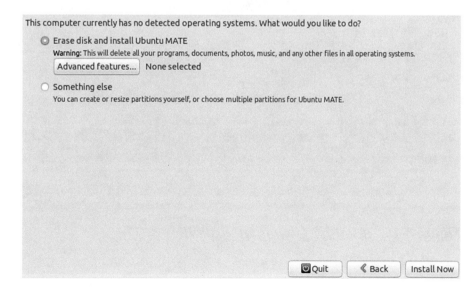

Figure 3-17. *Installing Ubuntu in VirtualBox*

17.   Click Continue. See Figure 3-18.

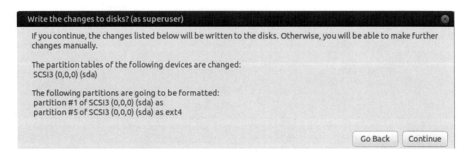

Figure 3-18. *Installing Ubuntu in VirtualBox*

18.   Select the region and click Continue again. See
Figure 3-19.

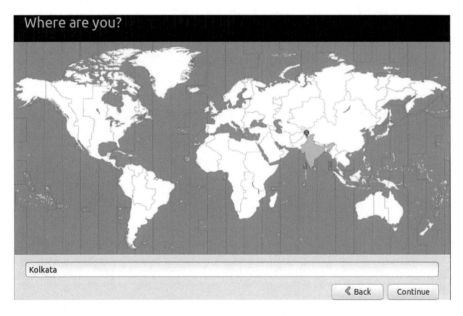

Figure 3-19. *Installing Ubuntu in VirtualBox*

19. Type your name, the computer name, your
     username, and a password and click Continue. See
     Figure 3-20.

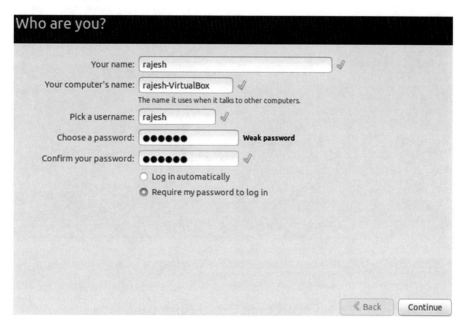

Figure 3-20. *Installing Ubuntu in VirtualBox*

20. Click Restart Now. See Figure 3-21.

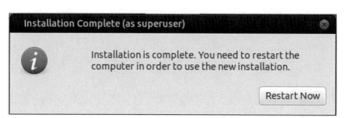

Figure 3-21. *Installing Ubuntu in VirtualBox*

21. From the Devices menu, select Optical Drives and
     then select the ISO filename. See Figure 3-22.

Figure 3-22. Installing Ubuntu in VirtualBox

22.   Select Remove Disk from Virtual Drive, as shown in
        Figure 3-23.

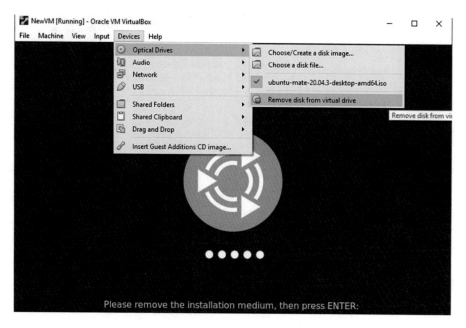

Figure 3-23. *Installing Ubuntu in VirtualBox*

23.   Click Force Unmount. See Figure 3-24.

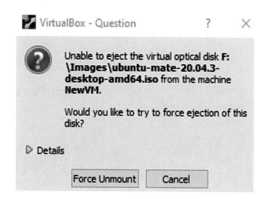

Figure 3-24. *Installing Ubuntu in VirtualBox*

24.   Click Install Now to install the updates. See
      Figure 3-25.

Figure 3-25. *Installing Ubuntu in VirtualBox*

25.   Click the Shut Down option. See Figure 3-26.

Figure 3-26.   *Installing Ubuntu in VirtualBox*

26.   Click Shut Down. See Figure 3-27.

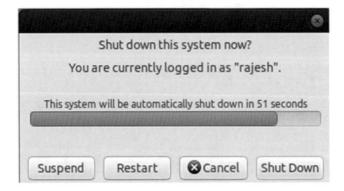

Figure 3-27. *Installing Ubuntu in VirtualBox*

27. Select the virtual machine (it's called NewVM in this example) from the left menu and click Settings. See Figure 3-28.

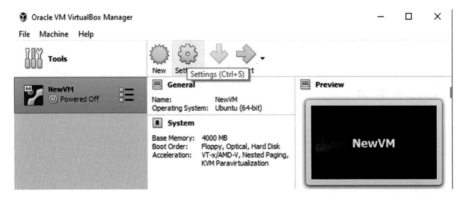

Figure 3-28. *Installing Ubuntu in VirtualBox*

28. Click Display and from the Screen option, set Video Memory to 128MB, and click OK. See Figure 3-29.

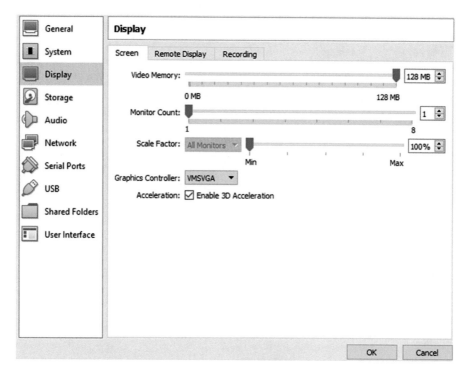

Figure 3-29. *Installing Ubuntu in VirtualBox*

29.   Select System and from the Processor tab, set the
      number of processor cores accordingly, and then
      click OK. See Figure 3-30.

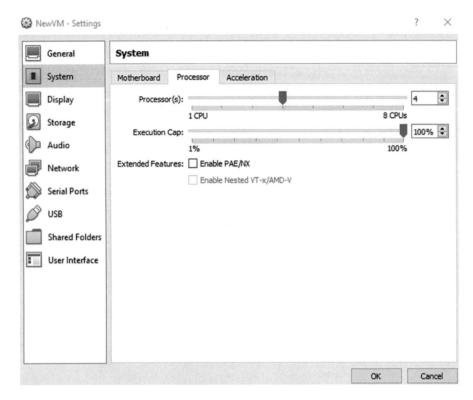

Figure 3-30. *Installing Ubuntu in VirtualBox*

Ubuntu installation and configuration in VirtualBox are now complete. The next section explains the ROS installation process.

ROS Noetic Installation

To install ROS Noetic, follow these instructions:

1. First, you need to configure Ubuntu OS to accept software from the `http://packages.ros.org/ros/ubuntu` website:

```
rajesh@rajesh-VirtualBox:~$ sudo sh -c 'echo "deb http://packages.ros.org/ros/ubuntu $(
lsb_release -sc) main" > /etc/apt/sources.list.d/ros-latest.list'
```

2. You need to install the curl software, which is used to transfer files between the workstation and the ROS webserver.

```
rajesh@rajesh-VirtualBox:~$ sudo apt install curl
```

3. Add a security key so that the system can make sure the software you install is genuinely created by Open Robotics (maintainers of ROS, and Gazebo):

```
rajesh@rajesh-VirtualBox:~$ curl -s https://raw.githubusercontent.com/ros/rosdistro/master/ros.asc | sudo apt-key add -
```

4. Get information about the latest version of the software and its dependencies using this command:

```
rajesh@rajesh-VirtualBox:~$ sudo apt update
```

5. Install ROS Noetic, rqt, RViz, the robot libraries, Gazebo, and so on.

```
rajesh@rajesh-VirtualBox:~$ sudo apt install ros-noetic-desktop-full
```

6. The source command executes a file called setup. bash located in the /opt/ros/noetic path. This loads some data values (environment variables) in the memory so that ROS can use them to work properly. You have to enter this command each time you open a new terminal:

```
rajesh@rajesh-VirtualBox:~$ source /opt/ros/noetic/setup.bash
```

7. You can permanently execute the `source` command to avoid entering the command each time you open a new terminal. This is done by writing the command into a file called `.bashrc`, which gets executed each time you open a new terminal.

```
rajesh@rajesh-VirtualBox:~$ echo "source /opt/ros/noetic/setup.bash" >> ~/.bashrc
```

```
rajesh@rajesh-VirtualBox:~$ source ~/.bashrc
```

8. There are a set of tools that you can use to conveniently manage workspaces and install the required software components (dependencies) in ROS. To install these tools, use the following commands:

```
rajesh@rajesh-VirtualBox:~$ sudo apt install python3-rosdep python3-rosinstall python3-rosinstall-generator python3-wstool build-essential
```

```
rajesh@rajesh-VirtualBox:~$ sudo apt install python3-rosdep
```

9. The `rosdep` tool allows automatic installation of the required dependencies of a ROS workspace. To use rosdep, you need to initialize it. That is done as follows:

```
rajesh@rajesh-VirtualBox:~$ sudo rosdep init
```

```
rajesh@rajesh-VirtualBox:~$ rosdep update
```

The ROS installation and configuration processes are complete. The next section explains how to install VS Code.

Visual Studio Code Installation

To install VS Code, follow these steps:

1. Download VS Code from this link:

    ```
    https://code.visualstudio.com/Download
    ```

    Refer to Figures 3-31 and 3-32.

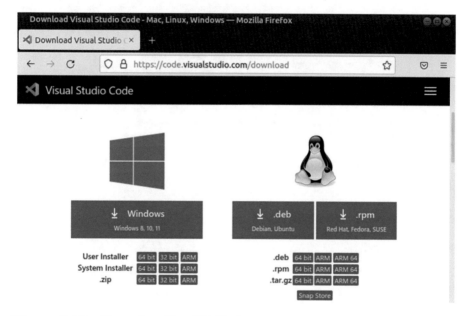

Figure 3-31. *Downloading VS Code*

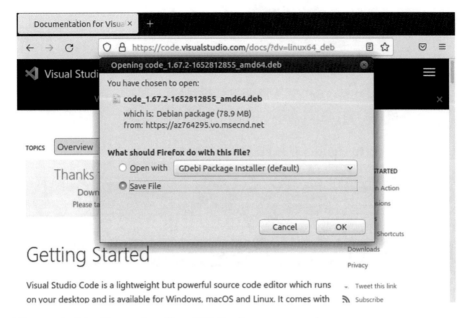

Figure 3-32. *Downloading VS Code*

2. Install the package.

   Install the downloaded file by making it executable
   (using the sudo chmod +x <filename> command,
   where <filename> is code_1.67.2-1652812855_amd64.
   deb). Then, double-click the file and install it. Refer to
   Figures 3-33, 3-34, and 3-35.

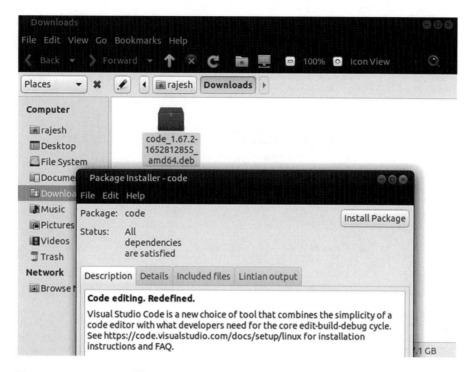

Figure 3-33. *Installing VS Code*

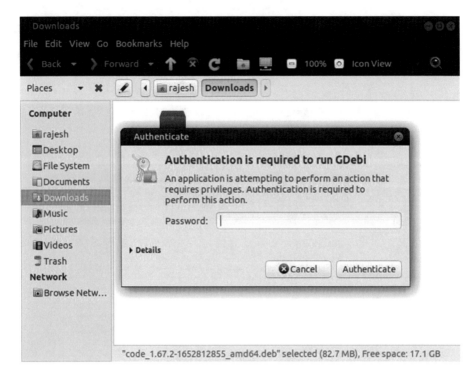

Figure 3-34. *Installing VS Code*

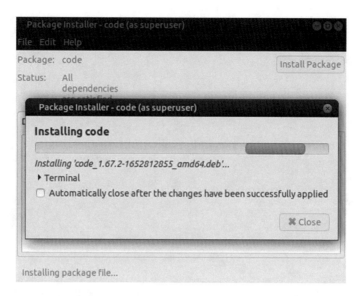

Figure 3-35. *Installing VS Code*

3. Run VS Code.

    To run VS Code, open a terminal and type the code
    command, as shown in Figure 3-36. Alternatively, you
    can search for VS Code in the menu (opened by pressing
    the Windows key on the keyboard).

Figure 3-36. *Executing VS Code*

4. Configure VS Code for ROS.

    VS Code has several plugins (called extensions) that
    provide additional features. To install these extensions,
    open VS Code and navigate to File ➤ Preferences ➤
    Extensions, as shown in Figure 3-37.

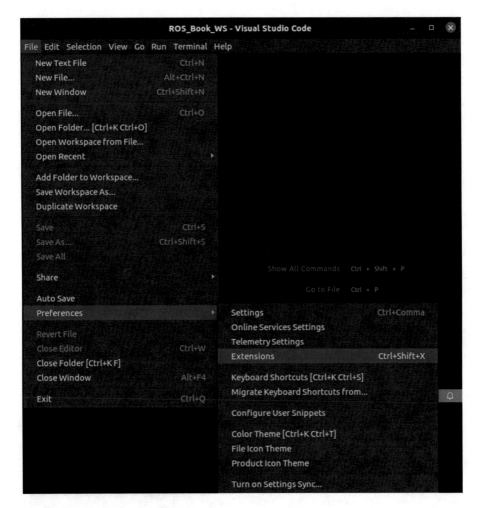

Figure 3-37. *VS Code Extensions menu*

VS Code Extensions

This section explains some of the VS Code extensions and their features.

ROS Extension

The ROS extension is used to manage ROS projects. This extension has the following features:

- ROS environment setup

- Start, terminate, and monitor the ROS master

- Perform compilation and building of ROS workspace and packages

- Create new ROS packages

- Run launch files and nodes

- Install dependencies of the workspace

- Syntax highlighting for ROS files

- Add the required C++/Python libraries automatically

- Preview the URDF robot model

- Debug ROS nodes

A screenshot of the extension is provided for reference in Figure 3-38.

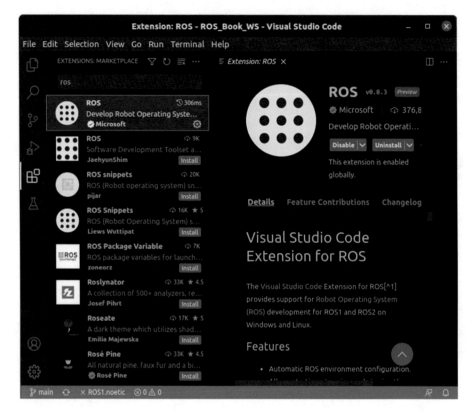

Figure 3-38. *ROS extension*

Python Extension

The Python extension assists you in writing scripts in the Python programming language. It has the following features:

- Auto-completion of Python scripts

- Code navigation

- Syntax checking

- Code analysis

- Code formatting

- Debugging

- Testing

- Virtual environment activation and switching

- Refactoring to improve the quality of the script

A screenshot of the extension is provided for reference in Figure 3-39.

Figure 3-39. *Python extension*

Pylint Extension

The Pylint extension assists you when writing code in the Python programming language. Its features are listed here:

- Lists errors in code

- Provides suggestions for a coding standard

- Refactors

A screenshot of the extension is provided for reference in Figure 3-40.

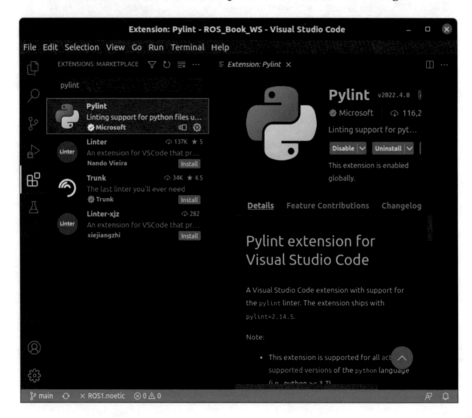

Figure 3-40. *Pylint extension*

CMake Extension

CMake is used to build ROS packages. Some of its features include:

- Syntax highlighting of CMake scripts

- Code completion

- Commenting in scripts

- Insert snippets (blocks of reusable code)

A screenshot of the extension is provided for reference in Figure 3-41.

Figure 3-41. *CMake extension*

C/C++ Extension

The C/C++ extension is used to assist you when writing code in C/C++ programming languages. This extension provides the following features:

- Syntax highlighting of codes

- Code completion

- Error checking

A screenshot of the extension is provided for reference in Figure 3-42.

Figure 3-42. *C/C++ extension*

VS Code installation and configuration are complete. The next section explains how to install a few other useful software programs.

Other Useful Software

Terminator: This is a terminal emulator (a window in which you enter commands) that can split the terminal window vertically and horizontally into multiple sub-sections. It is useful when you want to execute several ROS nodes or launch files and visualize their output and log messages. To install terminator, type the `sudo apt install terminator` command. A sample screen is shown in Figure 3-43.

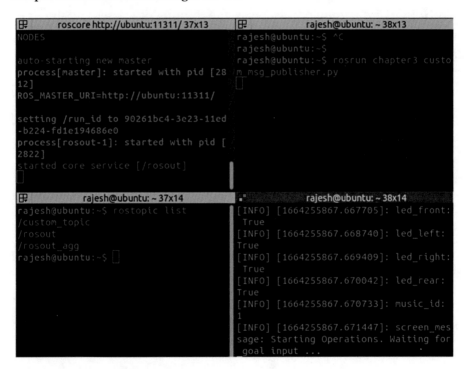

Figure 3-43. *Sample screenshot of Terminator*

GitKraken: This is a graphical software program for managing Git repositories. It allows you to track the repositories, visualizing branches/tags/commits, and so on. Download and install GitKraken from `www.gitkraken.com/download`.

Pinta: This is a drawing application that can be used to edit the maps generated by the robot. For example, the maps created by robots may contain noise data, people, dynamic objects, and so on, which can be removed by using this simple drawing tool. To install Pinta, go to Ubuntu Software and search for Pinta.

Blender: Blender is a free 3D modeling, animation, and rendering program. You can use it to create your robot model. To install Blender, go to Ubuntu Software and search for Blender.

Htop: This is a tool that visualizes the running processes and the resources they consume. To install Htop, type the `sudo apt install htop` command. A sample screenshot is provided in Figure 3-44.

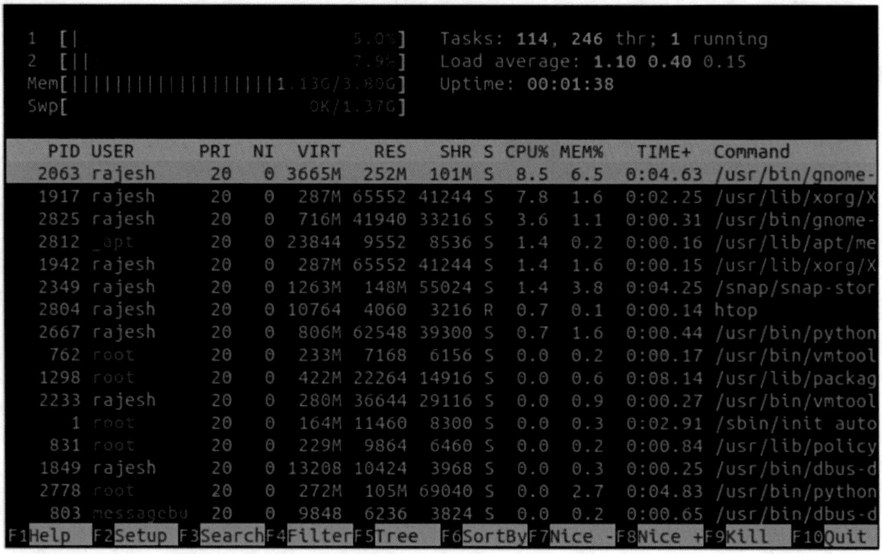

Figure 3-44. Htop sample screenshot

Summary

This chapter explained how to set up a workstation for simulation by installing Ubuntu, ROS, VS Code, and the other necessary software. The next chapter discusses ROS in detail.

CHAPTER 4

The ROS Framework

Outline

This chapter covers the following topics:

- What is ROS

- Why use ROS

- Applications of ROS

- ROS architecture

- Creating a workspace and building it

- Publishers/subscribers

- Services

- Actions

- Implementing publisher/subscriber using Python

- Implementing services using Python

- Implementing actions using Python

- Creating custom messages, custom service definitions, and custom actions

- Basic ROS, Catkin, and Git commands

© Rajesh Subramanian 2023
R. Subramanian, *Build Autonomous Mobile Robot from Scratch using ROS*,
Maker Innovations Series, https://doi.org/10.1007/978-1-4842-9645-5_4

- ROS debugging tools

- Coordinating transformations in ROS

- The ROS navigation stack

What Is ROS?

The Robot Operating System (ROS) is a platform used to create robots, both in simulation and in hardware. ROS includes a huge collection of prebuilt, modularized, customizable software packages that can be used as plugins in your robot. This allows the robot to perform various operations such as reading data from sensors, controlling the movements of actuators, mapping, localization, navigation, and more. ROS also provides a means to efficiently communicate between the various software components. Most of these prebuilt software packages are thoroughly tested, actively used by the robotics community all over the world, and frequently updated. All these features allow the roboticist to focus more on developing higher-level functionalities rather than reinventing the wheel by creating basic software packages for every robot.

Why Use ROS?

Suppose that you wanted to build a robot on your own, from scratch. That would typically require various software components:

- Drivers for the sensors and actuators

- PID control loop

- Forward/inverse kinematics

- Robot model definition

- Frame transformations

- Robot visualization

- Robot simulation

- Communication between different programs

- Motion planning, and so on

Creating all these software components from scratch, as well as testing and debugging them, would require a huge amount of time and effort. In addition, software developed for one robot could be quite challenging to reuse in another robot. Also, before you build your actual physical robot, you should test how the robot will behave in the real world. That would reduce time, cost, effort, and material waste.

ROS addresses all these issues by providing a generic robotic software framework that developers can use to create their robots easily. ROS has the following features:

- **Open-source**: Free to use and can be modified and distributed by the user.

- **Easy to use**: There is a huge collection of prebuilt packages that provide various functionalities.

- **Standard**: Certain conventions are followed while creating the software components (or packages), which results in a consistent and well-defined format. Users and other developers can easily modify existing packages or create new packages following those conventions, which results in improved efficiency.

- **Simulation**: Provides a built-in simulator and visualizer tool. Simulation helps you get an idea of the robot's behavior in the real world. Visualization helps you view the world from the robot's perspective.

- **Hardware flexibility**: Can run on almost any computational device, from single board computers such as Raspberry Pis to desktop/laptop computers.

- **Generic**: Can be used to build any type of robot, not just one particular type.

- **Modularized**: Has individual software components with standardized interfaces. This makes the software:

  - More readable

  - Reusable

  - More manageable

  - Easier to test and debug

- **Huge community and support**: Used worldwide by individuals, researchers, universities, industries, and so on. This allows users to find resources and get support from communities. Resources can be found in online courses, websites, blogs, textbooks, research papers, and so on. Support could be found on forums such as `https://answers.ros.org/`, `https://www.reddit.com/r/ROS/`, and so on.

Applications of ROS

ROS-integrated robots are used in areas such as industries, automotive, medical, logistics, warehouses, defense, security, agriculture, personal service, household, entertainment, and more. Some popular ROS-based robotics companies include ABB, FANUC, KUKA, Microsoft, Omron Corporation, Universal Robotics, Clearpath Robotics, Robotnik, iRobot, Rethink Robotics, and Husarion.

There are many robots that use ROS. A few well-known ROS-based robots are listed in Table 4-1.

Table 4-1. *ROS-Based Robots*

Robot	Category	Manufacturer
Turtlebot4	Indoor, ground	Clearpath Robotics
Panther	Outdoor, ground	Husarion
ARI	Humanoid	PAL Robotics
RB Vulcano	Mobile manipulator	Robotnik
UR Series	Industrial manipulator	Universal Robots
Ned2	Educational manipulator	Niryo
Heron USV	Marine, surface	Clearpath Robotics
BlueROV	Marine, underwater	Blue Robotics
Gapter	Aerial, educational, copter	Gaitech
Erle-Plane	Aerial, fixed wing	Erle Robotics

For more information about ROS-enabled robots, see `https://robots.ros.org/`.

The ROS Architecture

The ROS architecture refers to the underlying structure and concepts of ROS. The architecture of ROS has three parts:

- Filesystem

- Computation graph

- Community (resources and forums, sharing and collaborating resources, and knowledge with the ROS community worldwide)

The Filesystem

The *filesystem* refers to how files and folders are structured. The filesystem includes the following components:

- **Package**: This is the minimal software structure needed to build a feature. An example of a package could be to read data from a sensor (e.g., camera, Lidar etc.). The sensor package could read data from the sensor and send the sensor data in the form of messages that could be read by other programs.

- **Metapackage**: Contains a group of packages that bring about a higher functionality. For example, navigation, robotic arm movements, and so on.

- **Package manifest**: Within each package, there is a file that contains some information about the package, such as package name, version, description, maintainer, licenses, dependencies (such as libraries, messages, services), and so on. This file is called the package manifest and resides in the package as `package.xml`.

- **Metapackage manifest**: This is similar to a package manifest and contains information about the metapackage, such as metapackage name, licenses, maintainer, dependent package names, and so on. This file is also called `package.xml`.

- **Messages**: Communication between various components is performed via messages. There are several built-in message types in ROS (`http://wiki.ros.org/std_msgs`). You can also define custom messages, which are placed in the `msg` folder in the package.

- **Service**: This is a request-reply message used to communicate between programs. There are various built-in services in ROS (http://wiki.ros.org/ std_srvs). Custom services can also be defined and are placed in the srv folder within the package.

The ROS filesystem is depicted graphically in Figure 4-1.

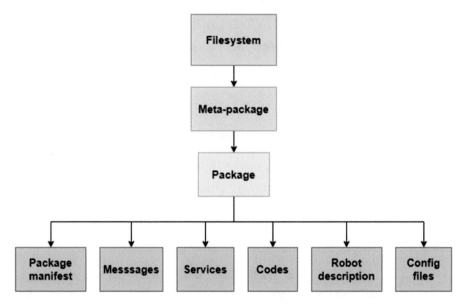

Figure 4-1. *The ROS filesystem*

If you want to create software using ROS for your robot, you start with a folder called workspace in which all the software components are placed. Inside the workspace folder, you can find three main folders: src, build, and devel.

- src: Contains packages, metapackages, and manifest files. Within the source directory, you place the components in various folders such as:

    - launch: Runs various programs of the package conveniently

    - config: Stores data values that could be accessed by ROS components

    - scripts: Writes scripts

    - src: Writes code

    - urdf: Describes robot joints and links

    - msg: Custom message definitions

    - srv: Custom service definitions

- build: Contains relevant files to build the packages

- devel: Contains the programs after compiling them

Note   Scripts are programs that can be directly executed without compilation. Code contains programs that need to be compiled before executing. It is a convention to place scripts in the scripts folder and code in the src folder. The scripts folder is a user-defined folder and is not generated by ROS. The src folder is automatically generated by ROS when you create a package with certain dependencies such as rospy or roscpp.

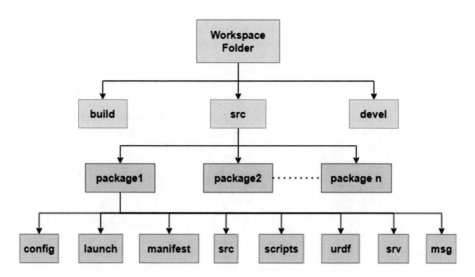

Figure 4-2. *The ROS workspace layout*

Figure 4-2 shows the general structure of a ROS workspace. It illustrates a real-world example of a ROS workspace. Figures 4-3 to 4-7 show the folder structure of a robot, where BumbleBot_WS is the name of the robot workspace. Within that workspace, you find the build, devel, src folders and other subfolders.

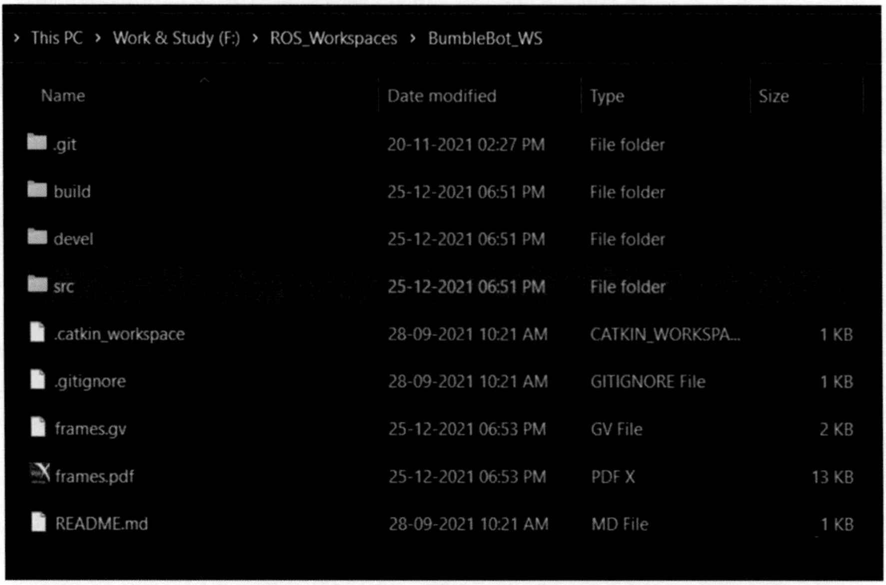

Figure 4-3. *Example of the ROS folder structure*

This PC > Work & Study (F:) > ROS_Workspaces > BumbleBot_WS > src			
Name	Date modified	Type	Size
arduino_codes	28-09-2021 10:21 AM	File folder	
bringup	28-09-2021 10:21 AM	File folder	
custom_scripts	28-09-2021 10:21 AM	File folder	
differential_driver	28-09-2021 10:21 AM	File folder	
learning_joy	28-09-2021 10:21 AM	File folder	
my_motor_controller	28-09-2021 10:21 AM	File folder	
my_robot_model	28-09-2021 10:21 AM	File folder	
resources	28-09-2021 10:21 AM	File folder	
robot_upstart	28-09-2021 10:21 AM	File folder	
rplidar_ros	28-09-2021 10:21 AM	File folder	
turtlebot_teleop	28-09-2021 10:21 AM	File folder	
BumbleBot.png	28-09-2021 10:21 AM	PNG File	524 KB
CMakeLists.txt	19-06-2022 05:57 PM	Text Document	1 KB

Figure 4-4. *Example of the ROS folder structure*

This PC > Work & Study (F:) > ROS_Workspaces > BumbleBot_WS > src > learning_joy			
Name	Date modified	Type	Size
launch	28-09-2021 10:21 AM	File folder	
src	28-09-2021 10:21 AM	File folder	
CMakeLists.txt	28-09-2021 10:21 AM	Text Document	7 KB
package.xml	28-09-2021 10:21 AM	XML File	3 KB

Figure 4-5. *Example of the ROS folder structure*

141

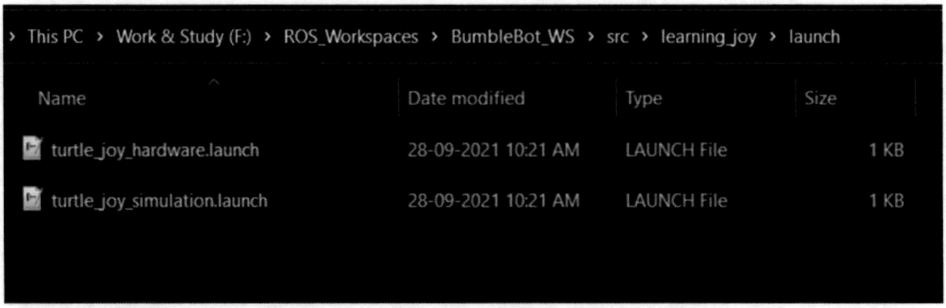

> This PC › Work & Study (F:) › ROS_Workspaces › BumbleBot_WS › src › learning_joy › launch

Name	Date modified	Type	Size
turtle_joy_hardware.launch	28-09-2021 10:21 AM	LAUNCH File	1 KB
turtle_joy_simulation.launch	28-09-2021 10:21 AM	LAUNCH File	1 KB

Figure 4-6. *Example of the ROS folder structure*

This PC › Work & Study (F:) › ROS_Workspaces › BumbleBot_WS › src › learning_joy › src

Name	Date modified	Type	Size
joy_node.py	28-09-2021 10:21 AM	PY File	1 KB
turtle_teleop_joy.cpp	28-09-2021 10:21 AM	CPP File	2 KB

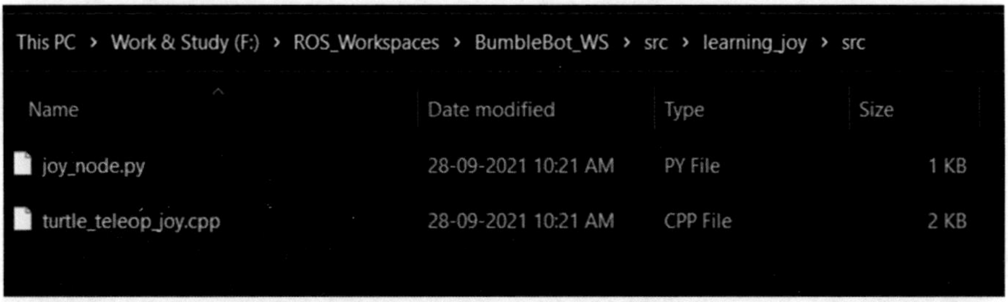

Figure 4-7. *Example of the ROS folder structure*

The Computation Graph

The *computation graph* describes how the programs communicate over a network. ROS has a network structure where the individual programs can communicate with other programs by sending and receiving data messages. The basic components of the computation graph are as follows:

- Node
- Master
- Messages and topics
- Publisher/subscriber
- Services

- Actions

- Bags

- Parameter server

A *node* is a program that provides a small functionality or part of a bigger functionality by communicating with other nodes. A typical robot will have several nodes connected to the ROS network. For example, a node receives distance values between the robot and surrounding obstacles and blinks when an obstacle is nearby. This node waits for the distance values sent from another node to blink the LED. Writing smaller nodes to attain a larger functionality is better than writing a single node. This provides better efficiency, robustness, ease of testing/debugging, and improves readability.

The *master* is a central node. Because ROS is a collection of separate nodes, you need a central node so the nodes can discover each other and communicate. The ROS master gathers information about the names, topics, and services offered by various nodes. For example, suppose that you have a Joystick node and a Move_Robot node. The Joystick node captures the button presses from the joystick and transmits the data. The Move_Robot node receives the joystick data and drives the robot according to the button presses. First, the Joystick node notifies itself to the master node and says it needs to publish (transmit) some data under the topic (name) called joystick_data (see Figure 4-8). Then, the Move_Robot node informs the master that it needs to subscribe (receive) data in the topic joystick_data (see Figure 4-9). The master now informs both the nodes about their presence, which enables the Joystick node and the Move_ Robot node to directly communicate with each other (see Figure 4-10).

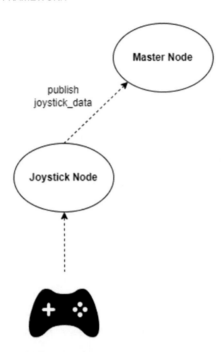

Figure 4-8. *The joystick node registering with the master node*

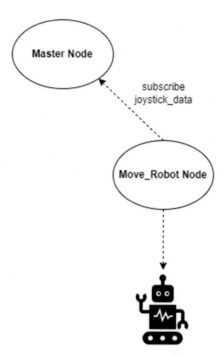

Figure 4-9. *The Move_Robot node registering with the master node*

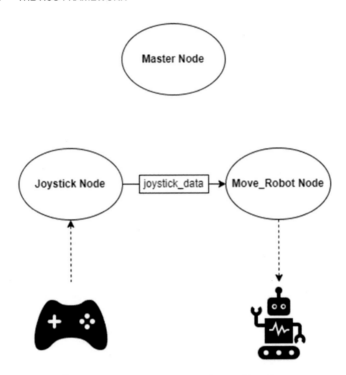

Figure 4-10. *Nodes communicating with each other*

Messages and topics: The exchange of information between various nodes is carried out via messages. A message contains a heading that the nodes can use to communicate. This heading is called a topic and must be unique to avoid conflict. For example, if you want to send an email, you add a subject name. Below the subject, you draft the contents to communicate. The subject name of the email is similar to the topic name of the message and the text in the email is similar to the data in the message. ROS has several standard built-in messages that can be used for your needs. Moreover, ROS allows defining custom messages, which can be used to communicate between nodes. To learn more about ROS standard messages, see `http://wiki.ros.org/std_msgs`.

A node can publish (send) and subscribe (receive) messages. A node that publishes a message is called a *publisher*. A node that subscribes to a message is called a *subscriber*. The publisher and subscriber nodes are independent and remain unaware of each other's presence until the master node informs them. The publisher and subscriber that want to communicate with each other must have the same message topic. A node can publish and subscribe to any number of topics.

A *service* is a method that is used to communicate between nodes. Here, messages are exchanged between nodes in a request-reply fashion. Services are used when you need a response after sending a message. The publisher-subscriber method does not have feedback. For example, you need to command a robot to move to a goal position and get a response from the robot if it reached the goal. In this situation, a service would be more appropriate than using a publisher/subscriber.

An *action* is another method of communication between nodes, which allows you get intermediate responses or feedback while executing a particular task. For instance, if you want to get in-between positions of a robot after commanding it to move to a goal position, the action method should be used. Services do not provide feedback while executing the task.

ROS bags store messages and play them back when required. Messages can be sensor values, motor positions, intermediate goal positions, and so on. You can use these recorded messages to fine-tune your robot, correct errors, replicate motor movements, and so on. For example, if you want to record some arm gestures in a humanoid robot, motor positions need to be published and the robot arms need to be moved manually. Meanwhile, you can record the messages (i.e., motor position values) in a bag file. Then you can play back the bag file to reproduce the arm gesture.

A *parameter server* is a central location for storing values. It allows all the ROS nodes to access those values globally. For example, say you want to specify the robot name, several wheels, acceleration, velocity, and so on, required by various nodes in the robot. You can store those values in the parameter server so that they could be accessed by the nodes. Moreover, those values can be configured to change the robot's behavior.

The ROS Community

ROS is used all over the world by researchers, industrialists, academicians, hobbyists, and so on. A huge number of resources and support exist in the form of the following:

- **Distributions**: A distribution is a stable version of a *metapackage* (a collection of software packages) that can be readily installed on a computer. Each distribution of ROS requires a particular version of Ubuntu to run smoothly. ROS has several distributions, as shown in Table 4-2.

Table 4-2. *ROS Distributions*

Distribution	Release Date	End of Life Date	Compatible Ubuntu Distribution
Noetic Ninjemys	23 May 2020	May 2025	Ubuntu Focal Fossa (20.04)
Melodic Morenia	23 May 2018	May 2023	Ubuntu Artful (17.10) Ubuntu Bionic (18.04)
Lunar Loggerhead	23 May 2017	May 2019	Ubuntu Xenial (16.04) Ubuntu Yakkety (16.10) Ubuntu Zesty (17.04)
Kinetic Kame	23 May 2016	April 2021	Ubuntu Wily (15.10) Ubuntu Xenial (16.04)
Jade Turtle	23 May 2015	May 2017	Ubuntu Trusty (14.04) Ubuntu Utopic (14.10) Ubuntu Vivid (15.04)

(continued)

Table 4-2. (*continued*)

Distribution	Release Date	End of Life Date	Compatible Ubuntu Distribution
Indigo Igloo	22 Jul 2014	April 2019	Ubuntu Saucy (13.10) Ubuntu Trusty (14.04 LTS)
Hydro Medusa	4 Sep 2013	May 2015	Ubuntu Precise (12.04 LTS) Ubuntu Quantal (12.10) Ubuntu Raring (13.04)
Groovy Galapagos	31 Dec 2012	July 2014	Ubuntu Oneiric (11.10) Ubuntu Precise (12.04 LTS) Ubuntu Quantal (12.10)
Fuerte Turtle	23 Apr 2012	--	Ubuntu Lucid (10.04 LTS) Ubuntu Oneiric (11.10) Ubuntu Precise (12.04 LTS)
Electric Emys	30 Aug 2011	--	Ubuntu Lucid (10.04 LTS) Ubuntu Maverick (10.10) Ubuntu Natty (11.04) Ubuntu Oneiric (11.10)
Diamondback	2 Mar 2011	--	Ubuntu Lucid (10.04 LTS) Ubuntu Maverick (10.10) Ubuntu Natty (11.04)
C Turtle	2 Aug 2010	--	Ubuntu Jaunty (9.04) Ubuntu Karmic (9.10) Ubuntu Lucid (10.04 LTS) Ubuntu Maverick (10.10)
Box Turtle	2 Mar 2010	--	Ubuntu Hardy (8.04 LTS) Ubuntu Intrepid (8.10) Ubuntu Jaunty (9.04) Ubuntu Karmic (9.10)

- **Repositories**: ROS packages are available in distributed
  storage spaces called *repositories*. Various organizations
  and individuals can develop their own ROS packages
  and release them into repositories. Most commonly
  used ROS packages are stored by many GitHub
  institutions. This enables other users to use those
  packages in their robots. For more information, see
  `http://wiki.ros.org/RecommendedRepositoryUsage/`
  `CommonGitHubOrganizations`.

- **ROS wiki**: This serves as the primary source of
  documentation of ROS. Information, tutorials, and
  other resources about various features, packages, and
  so on, are available on this page. It also allows users
  to create accounts, and write documentation and
  tutorials. For more information, see `http://wiki.`
  `ros.org/`.

- **Bug ticket system**: This is a feature that allows users to
  report issues, request new features, and so on, related
  to ROS. See `http://wiki.ros.org/Tickets`. To report
  an issue, follow these steps:

  - Open the Git page for the package.

  - Create/log in to your Git account.

  - Click the Issues tab.

  - Click New Issue.

  - Describe the issue.

  - Click the Submit the Issue button.

- **ROS Discourse**: This is a platform to inform users of the latest updates and news about ROS. See `https://discourse.ros.org/`.

- **ROS Answers**: This is a web forum for ROS where you can post queries and get support from other ROS users. See `https://answers.ros.org/`.

Note   ROS mailing lists and blogs have been replaced with ROS Discourse.

Creating a Workspace and Building It

To create and build a ROS workspace, follow these steps:

1. Create an empty folder. You can do this by right-clicking the mouse and clicking Create Folder or by typing the command shown in Figure 4-11.

```
rajesh@ROG-Zephyrus-G15:~$ mkdir ros_ws
```

Figure 4-11. *Creating a workspace folder*

2. List the folders using the `ls` command, as shown in Figure 4-12.

```
rajesh@ROG-Zephyrus-G15:~$ ls
Desktop     Downloads   Pictures    ros_ws    Templates
Documents   Music       Public      snap      Videos
rajesh@ROG-Zephyrus-G15:~$
```

Figure 4-12. *Listing the contents of the home directory*

3. Navigate to the ros_ws folder, as shown in
    Figure 4-13.

```
rajesh@ROG-Zephyrus-G15:~$ cd ros_ws
rajesh@ROG-Zephyrus-G15:~/ros_ws$
```

Figure 4-13. *Navigating to the ros_ws folder*

4. Create a folder named src inside the workspace, as
    shown in Figure 4-14.

```
rajesh@ROG-Zephyrus-G15:~/ros_ws$ mkdir src
rajesh@ROG-Zephyrus-G15:~/ros_ws$
```

Figure 4-14. *Creating the src folder inside the workspace*

5. Build the workspace, as shown in Figure 4-15.

```
rajesh@ROG-Zephyrus-G15:~/ros_ws$ catkin_make
```

Figure 4-15. *Building the workspace*

6. After you build your workspace, the message in
    Figure 4-16 is displayed.

```
-- Configuring done
-- Generating done
-- Build files have been written to: /home/rajesh/ros_ws/build
####
#### Running command: "make -j16 -l16" in "/home/rajesh/ros_ws/build"
####
rajesh@ROG-Zephyrus-G15:~/ros_ws$
```

Figure 4-16. *The workspace was built without any errors*

7. Now you can see two new folders—build and devel—in the workspace, as shown in Figure 4-17.

```
rajesh@ROG-Zephyrus-G15:~/ros_ws$ ls
build   devel   src
rajesh@ROG-Zephyrus-G15:~/ros_ws$
```

Figure 4-17. *Folder structure after building the workspace*

8. Navigate to the src folder, as shown in Figure 4-18.

```
rajesh@ROG-Zephyrus-G15:~/ros_ws$ cd src
rajesh@ROG-Zephyrus-G15:~/ros_ws/src$
```

Figure 4-18. *Navigating to the src folder*

9. Create a package called test_pkg with the rospy dependency using the command in Figure 4-19.

```
rajesh@ROG-Zephyrus-G15:~/ros_ws/src$ catkin_create_pkg test_pkg rospy
Created file test_pkg/package.xml
Created file test_pkg/CMakeLists.txt
Created folder test_pkg/src
Successfully created files in /home/rajesh/ros_ws/src/test_pkg. Please adjust th
e values in package.xml.
rajesh@ROG-Zephyrus-G15:~/ros_ws/src$
```

Figure 4-19. *Creating a new package named test_pkg*

10. Navigate to test_pkg, as shown in Figure 4-20.

```
rajesh@ROG-Zephyrus-G15:~/ros_ws/src$ ls
CMakeLists.txt   test_pkg
rajesh@ROG-Zephyrus-G15:~/ros_ws/src$ cd test_pkg/
rajesh@ROG-Zephyrus-G15:~/ros_ws/src/test_pkg$
```

Figure 4-20. *Navigating into the new package*

11.   Create a folder called `scripts` to store your Python scripts, as shown in Figure 4-21.

```
rajesh@ROG-Zephyrus-G15:~/ros_ws/src/test_pkg$ mkdir scripts
rajesh@ROG-Zephyrus-G15:~/ros_ws/src/test_pkg$
```

Figure 4-21.   *Creating a new folder named scripts*

12.   Create a Python script using the `gedit` text editor (or any other text editor), as shown in Figure 4-22.

```
rajesh@ROG-Zephyrus-G15:~/ros_ws/src/test_pkg/scripts$ gedit sample_script.py
```

Figure 4-22.   *Creating a Python script*

13.   The Python script is shown in Figure 4-23.

```
src > chapter3 > scripts >  sample_script.py
 1    #!/usr/bin/env python3
 2    '''Hello World Code'''
 3    print("Hello World !")
 4
```

Figure 4-23.   *Python script*

14.   Make the Python script executable, as shown in Figure 4-24.

```
rajesh@ROG-Zephyrus-G15:~/ros_ws/src/test_pkg/scripts$ sudo chmod +x sample_scri
pt.py
[sudo] password for rajesh:
```

Figure 4-24.   *Making the Python script executable*

15. After listing the files, you can see the Python script
     name displayed in green, which indicates that the
     script is executable. See Figure 4-25.

```
rajesh@ROG-Zephyrus-G15:~/ros_ws/src/test_pkg/scripts$ ls
sample_script.py
rajesh@ROG-Zephyrus-G15:~/ros_ws/src/test_pkg/scripts$
```

Figure 4-25. *Script changed to executable*

16. You need to source the workspace to run it using
     ROS. The source command helps you perform this
     operation. However, you have to run the source
     command each time when opening a new terminal.
     See Figure 4-26.

```
rajesh@ROG-Zephyrus-G15:~/ros_ws/src/test_pkg/scripts$ source ~/ros_ws/devel/set
up.bash
```

Figure 4-26. *Sourcing the workspace*

17. After sourcing, you can run the Python node using
     the rosrun command, as shown in Figure 4-27.

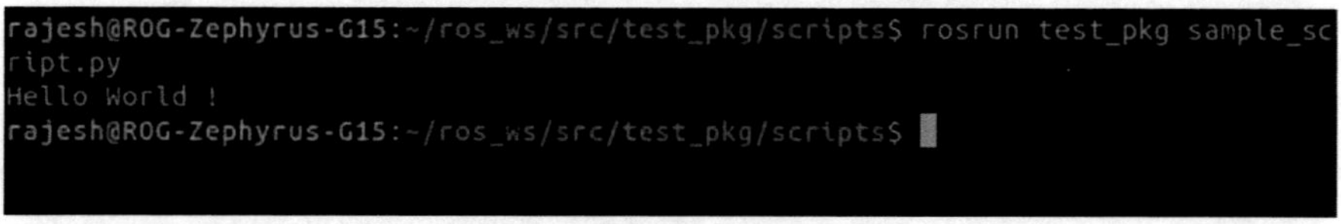

```
rajesh@ROG-Zephyrus-G15:~/ros_ws/src/test_pkg/scripts$ rosrun test_pkg sample_sc
ript.py
Hello World !
rajesh@ROG-Zephyrus-G15:~/ros_ws/src/test_pkg/scripts$
```

Figure 4-27. *Running the Python script*

18. You can also permanently source it by adding the source command to the .bashrc file. This avoids sourcing the workspace each time you open a new terminal. See Figure 4-28.

Figure 4-28. *Permanent sourcing by editing the .bashrc file*

Publishers/Subscribers

A publisher/subscriber process is a mechanism that allows different nodes to communicate with each other by sending/receiving messages. A node can publish messages, subscribe to messages, or do both. The publisher and subscriber nodes are independent and remain unaware of each other's presence until the master node informs them. The publisher and subscriber that want to communicate with each other must have the same message topic. A node can publish and subscribe to any number of topics. In Figure 4-29, you can see a publisher and subscriber communicating by exchanging messages over a topic.

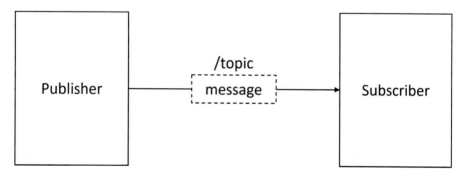

Figure 4-29. *Publisher/subscriber*

You can use the prebuilt message types provided by ROS or create your custom message type. The standard message types in ROS are as follows:

- Bool
- Byte
- ByteMultiArray
- Char
- ColorRGBA
- Duration

- Empty

- Float32

- Float32MultiArray

- Float64

- Float64MultiArray

- Header

- Int16

- Int16MultiArray

- Int32

- Int32MultiArray

- Int64

- Int64MultiArray

- Int8

- Int8MultiArray

- MultiArrayDimension

- MultiArrayLayout

- String

- Time

- UInt16

- UInt16MultiArray

- UInt32

- UInt32MultiArray

- UInt64

- UInt64MultiArray

- UInt8

- UInt8MultiArray

See http://wiki.ros.org/std_msgs for more information.

Services

A *service* is another method that is used to communicate between nodes. Here, the messages are exchanged between nodes in a request-reply manner. Services are used when you need a response after sending a message. The publisher-subscriber method does not have a response message. For example, say you needed to command your robot to move to a goal position and return a response when it reached the goal. In this situation, a service would be more appropriate than using a publisher/subscriber. Figure 4-30 shows a server and client communicating with each other using request and reply messages.

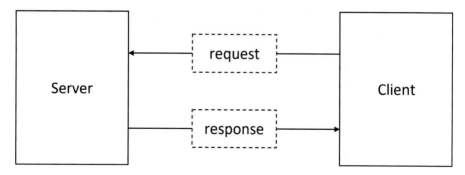

Figure 4-30. *Service server and service client*

Similar to messages in publisher/subscriber, you can use the standard service types in ROS or create your own service definitions. The built-in service types in ROS are as follows:

- `Empty`
- `SetBool`
- `Trigger`

See `http://wiki.ros.org/std_srvs` for more information.

Actions

An *action* is another method of communication between nodes. It allows you to get intermediate responses or feedback when executing a particular task. For instance, if you want to get intermediate positions of a robot after commanding it to move to a goal position, the action method could be used. Services do not provide feedback while executing the task.

Figure 4-31 illustrates an action server and action client communicating with each other. Here, you can see different messages such as `goal`, `cancel`, `status`, `result`, and `feedback`.

- The `goal` message is a request message sent from the client to the server to operate, such as moving a robot to a particular goal position.

- If the client needs to cancel the goal, it can send a `cancel` message to the server. The client can also get various response messages from the server such as `status`, `result`, and `feedback`.

- The `status` message is a number that corresponds to a state of execution of operation, such as `succeeded`, `aborted`, and so on. The complete list of status codes is listed in Table 4-3.

- The result message is sent only once by the server to the client after completing the operations. The message provides information to the client, such as the final robot position, after moving the robot to a list of goal positions.

- The feedback message is sent by the server to the client, informing it about the step-by-step updates about the operation. For instance, a feedback message may inform the client about the list of goal positions reached so far by a robot.

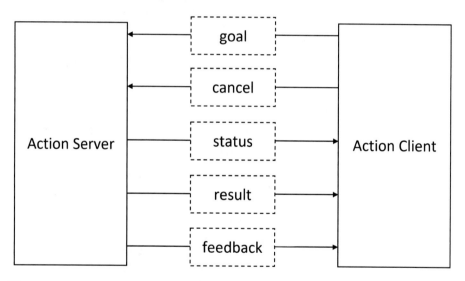

Figure 4-31. *Action server and action client*

Table 4-3. *Complete List of Status Codes*

Status Code	Status	Definition
0	PENDING	The action server has yet to execute operations.
1	ACTIVE	The action server is executing operations.
2	PREEMPTED	A cancel request was received by the action server while performing operations.
3	SUCCEEDED	Action server completed operations.
4	ABORTED	Action server aborted the operation because of a failure.
5	REJECTED	Action server rejected the operation because the goal could not be achieved or was invalid.
6	PREEMPTING	A cancel request was received by the action server while performing operations and has not finished execution.
7	RECALLING	The action server received a cancel request before operating, but the goal is cancellation was not confirmed by the action server so far.
8	RECALLED	The action server received a cancel request before it began operations and cancelled successfully.
9	LOST	Sometimes a goal can be lost, which can be identified by the action client.

See http://wiki.ros.org/actionlib for more information.

The next section explains how to implement the publisher, subscriber, service server, service client, action server, and action client. These operations are implemented using the Python3 programming language running on Ubuntu 20.04 operating system with ROS Noetic.

Implementing the Publisher/Subscriber Using Python

In this example, a publisher and a subscriber are implemented. They do the following:

- The publisher transmits a message.

- The subscriber listens to the message and displays it on the screen.

Publisher Code

```
1. #!/usr/bin/env python3
2. '''Publisher Code'''
3.
4. import rospy
5. from std_msgs.msg import String
6.
7. if __name__ == '__main__':
8.     try:
9.         pub = rospy.Publisher('pub_topic', String, queue_
            size=10)
10.        rospy.init_node('publisher_node')
11.        rate = rospy.Rate(1) # Frequency in Hz
12.        while not rospy.is_shutdown():
```

```
13.            pub.publish("hi !")
14.            rate.sleep()
15.    except rospy.ROSInterruptException as e:
16.        rospy.loginfo("Exception: " + str(e))
17.
```

Code Description

1. `#!/usr/bin/env python3`

The first line is called a "shebang." A shebang is used in scripting languages to specify which program should be used by the operating system to run the script. Here, it specifies that Python3 should be used to run this script. You can also manually run the Python script using the `/usr/bin/env python3 <filename.py>` command in the terminal.

4. `import rospy`

The line indicates that you are using a library called `rospy`, which contains some features. This library allows you to use various features of ROS in your programs, such as creating a publisher/subscriber, sending/receiving messages, and so on.

5. `from std_msgs.msg import String`

You also import another library called `std_msgs`, which contains a collection of predefined message types in ROS. Here, you use the `String` message type. The `String` message can store string data, such as `'hello'`, `'abc123'`, and so on. (See `http://wiki.ros.org/std_msgs`.)

7. `if __name__ == '__main__':`

This is where the program starts execution.

8.   `try:`

The `try:` statement handles errors that arise during the runtime of the program. Runtime errors (i.e., exceptions) can occur in a section of code within the `try` statement.

15. `except rospy.ROSInterruptException as e:`

The exceptions are handled by the `except` block, which prints a message. The `rospy.ROSInterruptException` is an exception that's raised when the program is terminated while executing.

16.   `rospy.loginfo("Exception: " + str(e))`

Displays any string message you intend to print on the screen.

9. `pub = rospy.Publisher('pub_topic', String, queue_size=10)`

This line creates a publisher with the following parameters:

- The topic name is the name of the topic under which you transmit messages. Here, `pub_topic` is the topic name.

- The message type is the type of message you want to send. It can be predefined message types such as `Bool`, `Int16`, `String`, and so on, or it can be custom message types. Here, `String` is the message type.

- Queue size specifies how many messages need to be stored before sending them. Here, `queue_size` is set to 10.

Note   If the rate of publishing messages is faster, you need to set a larger queue size. Similarly, if the rate of publishing messages is slower, then the queue size can be set to a smaller value.

If the queue size is smaller than what is required, some messages can be lost. On the contrary, if the queue size is larger than what is required, a lot of old messages will accumulate, which might cause the subscriber to process old data and could cause lag.

---

10. `rospy.init_node('publisher_node')`

This line initializes the node and provides a name you specify. The name must be unique, which means it should not be used by any other node in the system. Every node requires initialization, which enables it to communicate with the ROS master. Here, the name of the node is `publisher_node`. You can also provide an optional argument called `anonymous=True` when you initialize the node. This attaches a random number as a suffix to the node name, which ensures the node name is unique.

11. `rate = rospy.Rate(1) # Frequency in Hz`

The line `rate = rospy.Rate(1)` defines the rate (or frequency) of publishing messages. Here, the argument 1 means the frequency is one Hertz, that is, one message is transmitted per second. You can adjust the rate according to your requirements.

```
12. while not rospy.is_shutdown():
13.     pub.publish("hi !")
14.     rate.sleep()
```

This loop continuously publishes a string message called 'hi !' with a publishing frequency of 1Hz. The `rate.sleep()` method gives a time delay of one second between two consecutive messages. The time interval is automatically calculated using this formula:

Time interval between messages = 1/Frequency

The loop also checks whether the program is prompted to shut down using the `rospy.is_shutdown()` method.

Subscriber Code

```
1. #!/usr/bin/env python3
2. '''Subscriber Code'''
3. import rospy
4. from std_msgs.msg import String
5.
6. def callback(data):
7.     '''Callback function'''
8.     rospy.loginfo("Received Message- %s", data.data)
9.
10. if __name__ == '__main__':
11.     rospy.init_node('subscriber_node')
12.     rospy.Subscriber("pub_topic", String, callback)
13.     rospy.spin()
14.
```

Code Description

```
6. def callback(data):
```

The line defines a callback function. This method is executed when a message is received by the subscriber. The callback function displays the contents of the message.

12.   `rospy.Subscriber("pub_topic", String, callback)`

Here, a subscriber is created. The parameters are as follows:

- `pub_topic` is the topic name to listen to

- `String` is the data type of the message

- `callback` is the callback function

13.   `rospy.spin()`

`rospy.spin()` keeps the program running until termination. Without this line, the program would immediately stop after creating the subscriber.

Figures 4-32 and 4-33 show the output.

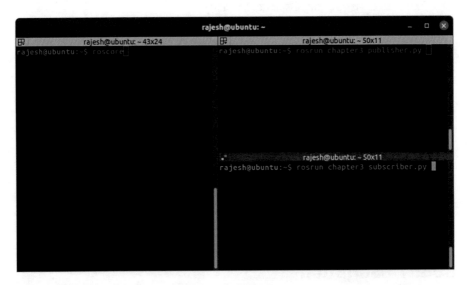

Figure 4-32. *Commands to run the ROS master, publisher, and subscriber*

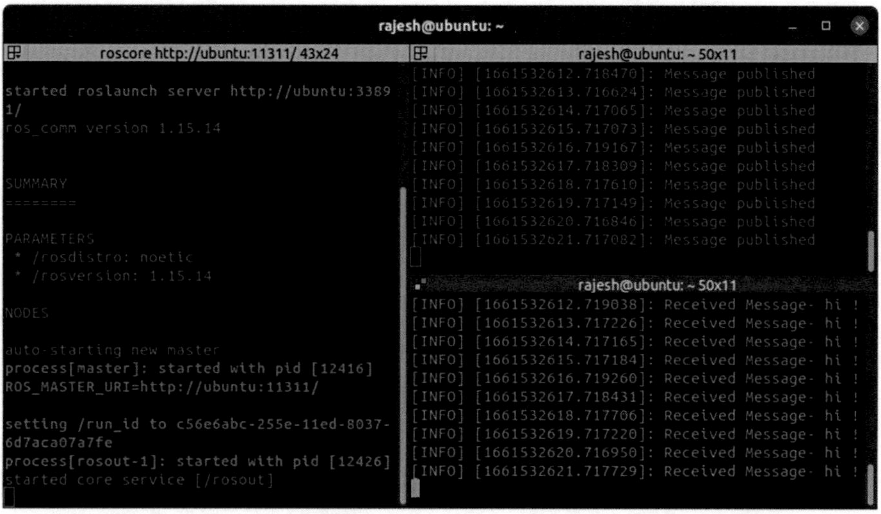

Figure 4-33. *ROS master, publisher, and subscriber running*

Implementing the Service Using Python

In this example, a service server/client are implemented. The following sections explain what they do.

Service Server

- The main aim of the server is to trigger a switch on/off according to the client's request.

- The server waits for the request message from the client.

- When a message is received, the contents are examined to determine what operation is to be performed on the switch.

- The server triggers the switch on/off accordingly. It also indicates that the operation was a success and sends an appropriate result message.

169

Service Client

- The client program prompts the user to enter input via the keyboard.

- Then the client sends a request message to the server indicating that the switch needs to be triggered ON/OFF.

- After sending the request, the client waits for the server operations to complete.

- When the reply message is received, the client displays the contents on the screen.

Service Server Code

```
1. #!/usr/bin/env python3
2. '''Service Server Code'''
3.
4. import rospy
5. from std_srvs.srv import SetBool, SetBoolResponse
6.
7. def handle_switch(req):
8.     '''Service handler'''
9.     rospy.loginfo("received request: " + str(req))
10.    reply_msg = SetBoolResponse()
11.    if req.data:
12.        reply_msg.success = True
13.        reply_msg.message = "Switch triggered ON"
14.    else:
15.        reply_msg.success = True
16.        reply_msg.message = "Switch triggered OFF"
17.    return reply_msg
```

```
18.
19. if __name__ == "__main__":
20.     rospy.init_node('trigger_switch_server')
21.     s = rospy.Service('/trigger_switch', SetBool,
        handle_switch)
22.     print("Ready to trigger switch")
23.     rospy.spin()
24.
```

Code Description

5. from std_srvs.srv import SetBool, SetBoolResponse

This example uses a predefined service message in ROS called
SetBool, which has three fields, as shown in Figure 4-34.

Figure 4-34. *SetBool service definition*

The data field is the request part. It is a true/false value (Boolean
type). The success and message fields constitute the reply part and are of
types Boolean and String, respectively.

The data field tells the server to turn on/off the switch. The success
field tells the client whether the operation was completed (i.e., the switch
was triggered on/off). The message field contains a message explaining the
operation in more detail to the client.

```
7. def handle_switch(req):
```

This is a method that handles service requests coming from the client. It performs the specified operations and sends a response message to the client.

```
10.     reply_msg = SetBoolResponse()
```

This line creates a response message.

```
11.     if req.data: # means, request is true
12.         reply_msg.success = True
13.         reply_msg.message = "Switch triggered ON"
14.     else: # means, request is false
15.         reply_msg.success = True
16.         reply_msg.message = "Switch triggered OFF"
17.     return reply_msg
```

This section determines whether the request data field is true or false. Then, it sets the success and message fields appropriately and sends a reply.

```
21.     s = rospy.Service('/trigger_switch', SetBool,
handle_switch)
```

This line creates a ROS service. The parameters are as follows:

- trigger_switch is the service name
- SetBool is the service type
- handle_switch is the callback function that processes the client request

Service Client Code

```
1. #!/usr/bin/env python3
2. '''Service Client Code'''
3.
4. import rospy
5. from std_srvs.srv import SetBool, SetBoolResponse
6.
7. if __name__ == '__main__':
8.     rospy.init_node("trigger_switch_client")
9.     rospy.wait_for_service("/trigger_switch")
10.
11.     try:
12.         response = SetBoolResponse()
13.         trigger_switch_service = rospy.ServiceProxy("/
            trigger_switch", SetBool)
14.         while True:
15.             trigger = input("Enter ON/OFF to trigger
                switch: ")
16.             if(trigger == "ON" or trigger == "on"):
17.                 response = trigger_switch_service(True)
                    #service call
18.                 break
19.             if(trigger == "OFF" or trigger == "off"):
20.                 response = trigger_switch_service(False)
                    #service call
21.                 break
22.             rospy.loginfo("Invalid input. Please try
                again.")
23.             continue
24.         rospy.loginfo("Response from Server:")
25.         rospy.loginfo("success- " + str(response.success))
```

```
26.            rospy.loginfo("message- " + str(response.message))
27.        except rospy.ServiceException as e:
28.            rospy.logwarn("Service failed: " + str(e))
29.            rospy.loginfo("Response from Server:")
30.            rospy.loginfo("success- " + "False")
31.            rospy.loginfo("message- " + "Exception !")
32.
```

Code Description

```
9.      rospy.wait_for_service("/trigger_switch")
```

In this line, the program will wait until the service becomes available. You can optionally provide a parameter named timeout, which waits a specific period (in seconds) for the service. If the client cannot connect to the server after the timeout period, an exception is raised.

```
13.            trigger_switch_service = rospy.ServiceProxy("/
               trigger_switch", SetBool)
```

Here, you provide the client with the details of the service.

- /trigger_switch is the service name

- SetBool is the service type

```
14.        while True:
15.            trigger = input("Enter ON/OFF to trigger
               switch: ")
16.            if(trigger == "ON" or trigger == "on"):
17.                response = trigger_switch_service(True)
                   #service call
18.                break
19.            if(trigger == "OFF" or trigger == "off"):
```

```
20.               response = trigger_switch_service(False)
                  #service call
21.               break
22.           rospy.loginfo("Invalid input. Please try
              again.")
23.           continue
```

Here, the user is prompted to enter a string input to turn a switch ON or OFF. If the input is on, then a true value is passed to the server. If the input is off, a false value is sent. If the user enters any string other than on/off, the 'Invalid input. Please try again.' message is displayed and the user is prompted to enter the input again.

```
25.           rospy.loginfo("success- " + str(response.success))
26.           rospy.loginfo("message- " + str(response.message))
```

These lines display success and response values, which are returned from the server after performing the operation.

Figures 4-35 through 4-37 show the output.

Figure 4-35. *Commands to run the ROS master, service server, and service client*

Figure 4-36. *Service client prompting the user to enter an ON/ OFF input*

Figure 4-37. *The server performs the operation and the client receives the reply*

Implementing Actions Using Python

In this example, an action server and an action client are implemented. The action server does the following:

- Here, the main goal of the action server is to wait for a list of goals from the client and move the robot to those goals sequentially.

- Once a goal list is received from the client, the server takes the goals one by one and waits for three seconds (assuming that the robot takes three seconds to reach each goal).

- As the robot moves, the server periodically sends feedback messages to the client. These messages contain the list of goals visited by the robot so far.

- If a cancel request is received from the client, the server cancels the robot operations. Also, the server returns a result message containing the partial list of visited goals.

- After performing operations, the server returns a status code and a message depending on the result.

The action client does the following:

- The client gets a list of goals from the user and sends it to the server.

- The client then receives periodic feedback from the server (goals covered by the robot so far) and displays it onscreen.

- The client waits for the server to finish operations, receive the status, and the result.

- Status and result are displayed onscreen.

- The client can also cancel the server operations if required. After the cancellation, the client waits for a partial result and status code from the server and displays it onscreen.

Action Server Code

```
1. #!/usr/bin/env python3
2. '''Action Server Code'''
3.
4. import time
5. import actionlib
6. import rospy
7. from chapter3.msg import MoveRobotAction
8. from chapter3.msg import MoveRobotFeedback
9. from chapter3.msg import MoveRobotResult
10. ACTION_SERVER = None
11.
12. def move_robot_callback(received_goals):
13.     ''' Callback '''
14.     global ACTION_SERVER
15.     reached_goals = []
16.     delay = 3 #wait time in seconds
17.     rate = rospy.Rate(1)
18.     rospy.loginfo("Goals have been received")
19.     rospy.loginfo(received_goals)
20.     success = False
21.     preempted = False
22.     i = 0
23.
```

```
24.     while not rospy.is_shutdown():
25.         #Taking each goal one by one and moving the robot
            towards it
26.         current_goal = received_goals.goals[i]
27.         rospy.loginfo("Moving to goal " + str(current_goal)
            + " ...")
28.         time.sleep(delay) #Dummy delay
29.         rospy.loginfo("Reached goal " + str(current_goal))
30.         reached_goals.append(current_goal)
31.         i = i+1
32.         if ACTION_SERVER.is_preempt_requested():
33.             preempted = True
34.             break
35.         if reached_goals == received_goals.goals:
36.             success = True
37.             break
38.         feedback_msg = MoveRobotFeedback()
39.         feedback_msg.visited_goals_history = reached_goals
40.         ACTION_SERVER.publish_feedback(feedback_msg)
41.         rate.sleep()
42.
43.     result = MoveRobotResult()
44.     if success:
45.         result.result = "All goals visited !"
46.     else:
47.         result.result = "All goals not visited ..."
48.     rospy.loginfo("sent goal result to client")
49.
50.     if preempted:
51.         rospy.loginfo("Preempted")
52.         ACTION_SERVER.set_preempted(result)
```

```
53.    elif success:
54.        rospy.loginfo("Success")
55.        ACTION_SERVER.set_succeeded(result)
56.    else:
57.        rospy.loginfo("Failure")
58.        ACTION_SERVER.set_aborted(result)
59.
60. def main():
61.    '''Main'''
62.    global ACTION_SERVER
63.    ACTION_SERVER = actionlib.SimpleActionServer('/move_
       robot', MoveRobotAction,
64.        execute_cb=move_robot_callback, auto_start=False)
65.    ACTION_SERVER.start()
66.
67. if __name__ == '__main__':
68.    rospy.init_node('move_robot_action_server')
69.    main()
70.    rospy.spin()
71.
```

Code Description

```
5. import actionlib
6. import rospy
7. from chapter3.msg import MoveRobotAction
8. from chapter3.msg import MoveRobotFeedback
9. from chapter3.msg import MoveRobotResult
```

This code uses actionlib, rospy, and a custom action message (Move_Robot) for this example.

The `actionlib` library allows you to perform several operations, such as:

- Create the action server/client

- Call the server

- Monitor feedback

- Get results

- Get the status

- Cancel goals

The `Move_Robot` action message contains the three message fields depicted in Figure 4-38.

goals
result
goals_reached_till_now

Figure 4-38. *Custom action message*

- `goals`: This is the list of goals to be traversed by the robot. It is passed from the client to the server.

- `result`: This is the information sent from the server to the client after performing the operations.

- `goals_reached_till_now`: This is the update message sent by the server to the client informing it about the goals visited by the robot up to now.

```
12. def move_robot_callback(received_goals):
```

This is a callback function that's invoked when a goal is received from the client.

```
24.       while not rospy.is_shutdown():
25.           #Taking each goal one by one and moving the robot
              towards it
26.           current_goal = received_goals.goals[i]
27.           rospy.loginfo("Moving to goal " + str(current_goal)
              + " ...")
28.           time.sleep(delay) #Dummy delay
29.           rospy.loginfo("Reached goal " + str(current_goal))
30.           reached_goals.append(current_goal)
31.           i = i+1
32.           if ACTION_SERVER.is_preempt_requested():
33.               preempted = True
34.               break
35.           if reached_goals == received_goals.goals:
36.               success = True
37.               break
38.           feedback_msg = MoveRobotFeedback()
39.           feedback_msg.visited_goals_history = reached_goals
40.           ACTION_SERVER.publish_feedback(feedback_msg)
41.           rate.sleep()
42.
43.       result = MoveRobotResult()
44.       if success:
45.           result.result = "All goals visited !"
46.       else:
47.           result.result = "All goals not visited ..."
48.       rospy.loginfo("sent goal result to client")
```

```
49.
50.    if preempted:
51.        rospy.loginfo("Preempted")
52.        ACTION_SERVER.set_preempted(result)
53.    elif success:
54.        rospy.loginfo("Success")
55.        ACTION_SERVER.set_succeeded(result)
56.    else:
57.        rospy.loginfo("Failure")
58.        ACTION_SERVER.set_aborted(result)
```

The goal positions are taken one by one from the list, and the program sleeps for a dummy time (here, three seconds). This sleep time corresponds to the time it takes the robot to reach the goal. After reaching the goal, the list of visited goals is updated. This list is sent as a feedback message to the client informing it about the progress of the operation.

- When the robot has visited all the goals, a result message 'All goals visited !' and a status code (3 for SUCCESS) is sent to the client.

- If at any point during the operation, the client cancels the goal (preempted), a 'All goals not visited ...' message and a status code (2 for PREEMPTED) are sent.

- If the server fails, a 'All goals not visited ...' message and a status code (corresponding to the state of the server) is sent.

```
63. ACTION_SERVER = actionlib.SimpleActionServer('/move_robot',
MoveRobotAction,
64.         execute_cb=move_robot_callback, auto_start=False)
```

This code creates an action server and specifies certain parameters for it.

- /move_robot is the action name

- MoveRobotAction is the action type

- move_robot_callback is the callback function that's invoked when a goal is received from the client

- auto_start=False invokes the action server at a later point in the code

65.   ACTION_SERVER.start()

On this line, you start the action server.

Action Client Code

```
1. #!/usr/bin/env python3
2. '''Action Client Code'''
3.
4. import rospy
5. import actionlib
6. from chapter3.msg import MoveRobotAction
7. from chapter3.msg import MoveRobotGoal
8.
9. def done_callback(status, result):
10.     '''Completed Callback'''
11.     if(status==0):
12.         rospy.loginfo("Status: " + "PENDING")
13.     if(status==1):
14.         rospy.loginfo("Status: " + "ACTIVE")
15.     if(status==2):
16.         rospy.loginfo("Status: " + "PREEMPTED")
```

```
17.     if(status==3):
18.         rospy.loginfo("Status: " + "SUCCEEDED")
19.     if(status==4):
20.         rospy.loginfo("Status: " + "ABORTED")
21.     if(status==5):
22.         rospy.loginfo("Status: " + "REJECTED")
23.     if(status==6):
24.         rospy.loginfo("Status: " + "PREEMPTING")
25.     if(status==7):
26.         rospy.loginfo("Status: " + "RECALLING")
27.     if(status==8):
28.         rospy.loginfo("Status: " + "RECALLED")
29.     if(status==9):
30.         rospy.loginfo("Status: " + "LOST")
31.
32.     rospy.loginfo("Result: " + str(result))
33.
34. def feedback_callback(feedback):
35.     '''Status Callback'''
36.     rospy.loginfo("Feedback: " + str(feedback))
37.
38. def main(action_client):
39.     '''Main'''
40.     while(True):
41.         goals_list = input("\nEnter the goal ID/s separated
            by commas. Eg- 1,2,3 : ")
42.         if(len(goals_list)>0):
43.             goals_list = goals_list.strip().split(',')
44.             goal = MoveRobotGoal(goals=goals_list)
45.             action_client.send_goal(goal, done_cb=done_
                callback, feedback_cb=feedback_callback)
```

```
46.               rospy.loginfo("Calling the action server")
47.               break
48.          else:
49.              continue
50. if __name__ == '__main__':
51.      rospy.init_node('move_robot_action_client')
52.      ACTION_CLIENT = actionlib.SimpleActionClient('/move_
         robot', MoveRobotAction)
53.      ACTION_CLIENT.wait_for_server()
54.      rospy.loginfo("Robot waiting for goals ...")
55.      main(ACTION_CLIENT)
56.      rospy.spin()
57.
```

Code Description

```
9. def done_callback(status, result):
```

This is the callback function that's invoked when the action server finishes its operations or is cancelled or fails due to some other reason. Depending on the status, a message is displayed on the screen. Also, the result message is displayed.

```
34. def feedback_callback(feedback):
35.      '''Status Callback'''
36.      rospy.loginfo("Feedback: " + str(feedback))
```

Here, the callback function is invoked upon receiving feedback messages from the server. The feedback message is then displayed on the screen.

```
40.     while(True):
41.         goals_list = input("\nEnter the goal ID/s separated
            by commas. Eg- 1,2,3 : ")
42.         if(len(goals_list)>0):
43.             goals_list = goals_list.strip().split(',')
44.             goal = MoveRobotGoal(goals=goals_list)
45.             action_client.send_goal(goal, done_cb=done_
                callback, feedback_cb=feedback_callback)
46.             rospy.loginfo("Calling the action server")
47.             break
48.         else:
49.             continue
```

In this section, the client gets input from the user through a keyboard. The input includes a set of goal IDs separated by commas (e.g. 1,2,3,4,5). The IDs are extracted (using strip and split functions) and added to a list. After that, the goal list is passed to the action server. If the user does not enter at least one goal ID, the input is prompted again.

```
52. ACTION_CLIENT = actionlib.SimpleActionClient('/move_robot',
MoveRobotAction)
```

This line creates an action client with some parameters:

- /move_robot is the action name

- MoveRobotAction is the action type

```
53.     ACTION_CLIENT.wait_for_server()
```

Here, the client waits for the action server to become available.
The output is shown in Figures 4-39 through 4-41.

Figure 4-39. *Commands to run the ROS master, action server, and action client*

Figure 4-40. *Action client prompting the user to enter the list of goals*

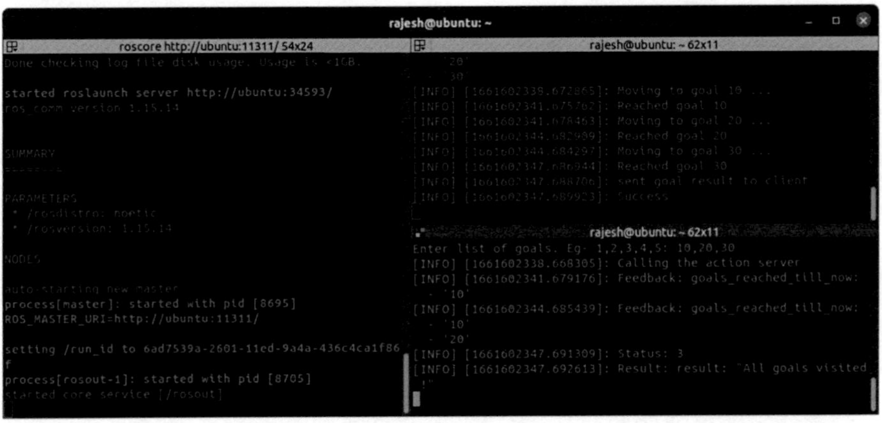

Figure 4-41. Action server performing the operations and client receiving feedback, status, and results

Creating Custom Messages, Custom Service Definitions, and Custom Actions

Even though ROS provides you with a set of predefined messages, services, and action definitions, you might want to define your own messages, services, and actions for convenience. ROS allows you to perform such operations.

Create a Custom Message Definition

A *message definition* is a plain text file that can contain one or more fields. Each field has a type and name. The types of fields that can be used in the message definition are as follows:

- Primitive types

  - `bool`

  - `int8`

  - `int16`

189

- int32

- int64

- uint8

- uint16

- uint32

- uint64

- float32

- float64

- string

- time

- duration

- Arrays (i.e., list of primitive types, such as int8[ ], bool[ ], etc.)

- Other messages (e.g., messages provided by the std_ msgs package; see http://wiki.ros.org/std_msgs for more information.)

To learn more about primitive datatypes in ROS, see http://wiki. ros.org/msg.

An example of a predefined message in ROS is the Twist message, which is provided by the geometry_msgs library. You can see the contents of this message in Figure 4-42.

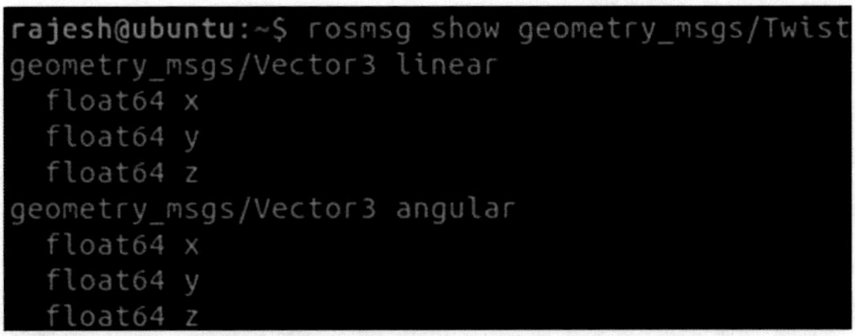

```
rajesh@ubuntu:~$ rosmsg show geometry_msgs/Twist
geometry_msgs/Vector3 linear
  float64 x
  float64 y
  float64 z
geometry_msgs/Vector3 angular
  float64 x
  float64 y
  float64 z
```

Figure 4-42. *Definition of the Twist message*

The Twist message provides linear and angular velocity commands to the robot. In this message, you can see another predefined message type called Vector3. Vector3 has three components—x, y, and z—which are of primitive type float64.

Procedure

You are about to create a custom message to control the LEDs, music, and text to display on screen. Follow these steps:

1. Navigate to the package, as shown in Figure 4-43.

```
rajesh@ubuntu:~$ roscd chapter3
rajesh@ubuntu:~/ROS_Book_WS/src/chapter3$
```

Figure 4-43. *Navigating to the package*

2. Create a folder called msg, as shown in Figure 4-44.

```
rajesh@ubuntu:~/ROS_Book_WS/src/chapter3$ mkdir msg
```

Figure 4-44. *Creating a message folder*

3. Within the folder, create a new file called <message_
   name>.msg, where <message_name> is the name of
   the message you intend to create. Create a custom
   message called robot_peripherals, as shown in
   Figure 4-45.

```
rajesh@ubuntu:~/ROS_Book_WS/src/chapter3/msg$ gedit robot_peripherals.msg
```

Figure 4-45. *Creating a message definition*

4. Define the custom message fields as shown in
   Figure 4-46 and then save the file.

Figure 4-46. *Describing a custom message*

5. Open the package.xml file located in the package
   and add the lines highlighted in yellow in
   Figure 4-47. This will convert the message definition
   to code during compilation. This code can be used
   in your ROS nodes.

```
<!-- Use doc_depend for packages you need only for
building documentation: -->
  <!--    <doc_depend>doxygen</doc_depend> -->
  <buildtool_depend>catkin</buildtool_depend>
  <build_depend>roscpp</build_depend>
  <build_depend>rospy</build_depend>
  <build_depend>std_msgs</build_depend>
  <build_depend>message_generation</build_depend>
  <build_depend>actionlib_msgs</build_depend>
  <build_export_depend>roscpp</build_export_depend>
  <build_export_depend>rospy</build_export_depend>
  <build_export_depend>std_msgs</build_export_depend>
  <build_export_depend>actionlib_msgs</-
build_export_depend>
  <exec_depend>roscpp</exec_depend>
  <exec_depend>rospy</exec_depend>
  <exec_depend>std_msgs</exec_depend>
  <exec_depend>actionlib_msgs</exec_depend>
  <exec_depend>message_runtime</exec_depend>
```

Figure 4-47.   *Editing the package.xml file*

6.   Open the CMakeLists.txt file located in the package and add the message_generation and std_msgs lines, as shown in Figure 4-48. This helps generate the custom message.

```
find_package(catkin REQUIRED COMPONENTS
  roscpp
  rospy
  std_msgs
  actionlib_msgs
  message_generation
)
```

Figure 4-48.   *Editing the CMakeLists.txt file*

7. In CMakeLists.txt, add the message_runtime and std_msgs lines, which are required during runtime. See Figure 4-49.

```
## DEPENDS: system dependencies of this project that
dependent projects also need
catkin_package(
#   INCLUDE_DIRS include
#   LIBRARIES chapter3
    CATKIN_DEPENDS roscpp rospy std_msgs actionlib_msgs
message_runtime
#   DEPENDS system_lib
)
```

Figure 4-49. *Editing the CMakeLists.txt file*

8. Also within the CMakeLists.txt file, add the section shown in Figure 4-50.

```
## Generate messages in the 'msg' folder
 add_message_files(
   FILES
   robot_peripherals.msg
 )
```

Figure 4-50. *Editing the CMakeLists.txt file*

9. You need to specify the required components
    (dependencies) by adding std_msgs to the
    CMakeLists.txt file, as shown in Figure 4-51.

```
## Generate added messages and services with any
dependencies listed here
generate_messages(
    DEPENDENCIES
    std_msgs
    actionlib_msgs
)
```

Figure 4-51. *Editing the CMakeLists.txt file*

10. Build the workspace using the catkin_make or
    catkin build command.

11. Check if the message has been generated. See
    Figure 4-52.

```
rajesh@ubuntu:~$ rosmsg show robot_peripherals
[chapter3/robot_peripherals]:
bool led_front
bool led_left
bool led_right
bool led_rear
int8 music_id
string screen_message
```

Figure 4-52. *Fields of the custom message*

In Figure 4-52, you can see that the custom message has
six fields. They are as follows:

- led_front: Toggles the front LED

- led_left: Toggles the left LED

- led_right: Toggles the right LED

- led_rear: Toggles the rear LED

- music_id: The ID of the music file to play

- screen_message: Text message to display on the screen of the robot

This next section explains how to publish and subscribe the custom message in your ROS nodes.

Publishing the Custom Message

```
1. #!/usr/bin/env python3
2. '''Publisher Code'''
3.
4. import rospy
5. from chapter3.msg import robot_peripherals
6.
7. if __name__ == '__main__':
8.     try:
9.         pub = rospy.Publisher('/custom_topic', robot_
           peripherals, queue_size=10)
10.        custom_msg = robot_peripherals()
11.        rospy.init_node('publisher_node')
12.        rate = rospy.Rate(1) # Frequency in Hz
13.        while not rospy.is_shutdown():
14.            custom_msg.led_front = True
15.            custom_msg.led_left = True
16.            custom_msg.led_right = True
17.            custom_msg.led_rear = True
18.            custom_msg.music_id = 1
```

```
19.              custom_msg.screen_message = "Starting
                 Operations. Waiting for goal input ..."
20.              pub.publish(custom_msg)
21.              rate.sleep()
22.      except rospy.ROSInterruptException as e:
23.          rospy.loginfo("Exception: " + str(e))
24.
```

This code publishes data using the custom message format. Line 5 imports the custom message. Line 10 uses the custom message type robot_peripherals to create a message variable. Lines 14-19 fill the fields of the message accordingly. On Line 20, the message is published.

Subscribing to the Custom Message

```
1. #!/usr/bin/env python3
2. '''Subscriber Code'''
3. import rospy
4. from chapter3.msg import robot_peripherals
5.
6. def callback(data):
7.     '''Callback function'''
8.     rospy.loginfo("Received Message")
9.     rospy.loginfo("led_front: %s", data.led_front)
10.    rospy.loginfo("led_left: %s", data.led_left)
11.    rospy.loginfo("led_right: %s", data.led_right)
12.    rospy.loginfo("led_rear: %s", data.led_rear)
13.    rospy.loginfo("music_id: %s", data.music_id)
14.    rospy.loginfo("screen_message: %s", data.screen_
       message)
15.
16. if __name__ == '__main__':
```

```
17.      rospy.init_node('subscriber_node')
18.      rospy.Subscriber("/custom_topic", robot_peripherals,
         callback)
19.      rospy.spin()
20.
```

This code subscribes data in the custom message format. Line 4 imports the custom message. Line 6 defines a callback function that is activated when a message is received. Then you extract each field in the message and display them on Lines 9-14.

The output is shown in Figures 4-53 and 4-54.

Figure 4-53. *Commands to run the publisher and subscriber*

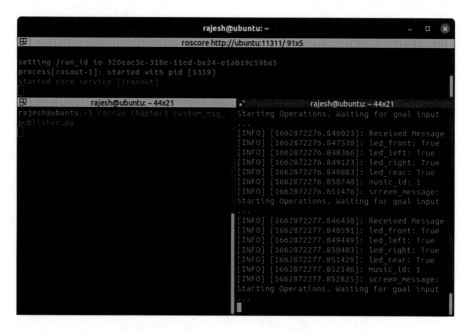

Figure 4-54. *Publisher and subscriber operating using the custom message*

Create a Custom Service Definition

A *service definition* is a plain text file that contains two parts—a request part and a reply part—separated by three hyphens (- - -). These parts can contain one or more fields and each field has a type and name. The types used in a service definition are as follows:

- Primitive types

    - bool

    - int8

    - int16

    - int32

    - int64

199

- uint8

- uint16

- uint32

- uint64

- float32

- float64

- string

- time

- duration

- Arrays (i.e., list of primitive types such as int8[ ], bool[ ], etc.)

- Other messages (for example, messages provided by the std_msgs package, custom messages, etc.)

To learn more about primitive datatypes in ROS, see http://wiki.ros.org/msg.

An example of a predefined service in ROS is the SetBool message, provided by the std_srvs library. You can see the contents of this service in Figure 4-55.

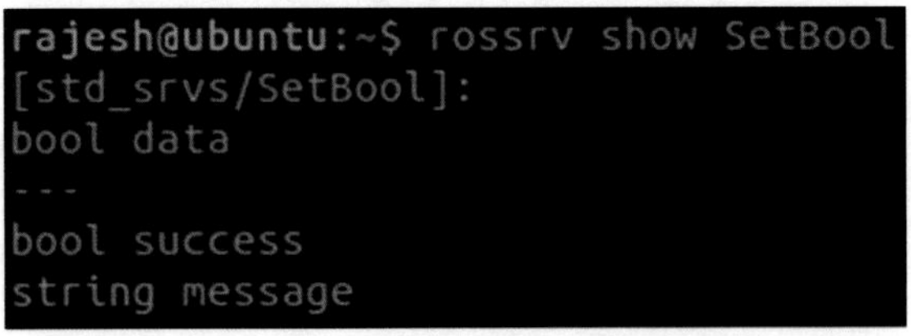

```
rajesh@ubuntu:~$ rossrv show SetBool
[std_srvs/SetBool]:
bool data
---
bool success
string message
```

Figure 4-55. *Definition of the SetBool message*

The SetBool service is typically used to trigger functionalities in a robot. In this message, there are two parts separated by the --- symbol. This symbol indicates the separation between request and response message parts. The top part is the request and the bottom part is the response. The field types used in this service are bool and string, which are primitive message types.

Procedure

Create a custom service to control the LEDs, music, and text to display onscreen and get a response. Follow these steps:

1. Navigate to the package, as shown in Figure 4-56.

```
rajesh@ubuntu:~$ roscd chapter3
rajesh@ubuntu:~/ROS_Book_WS/src/chapter3$
```

Figure 4-56. *Navigating to a package*

2. Create a folder named srv, as shown in Figure 4-57.

```
rajesh@ubuntu:~/ROS_Book_WS/src/chapter3$ mkdir srv
```

Figure 4-57. *Creating a service folder*

3. Within the srv folder, create a new file called
    <service_name>.srv, where <service _name> is the
    name of the service you intend to create. Here, you
    create a custom service called robot_accessories,
    as shown in Figure 4-58.

```
rajesh@ubuntu:~/ROS_Book_WS/src/chapter3/srv$ gedit robot_accessories.srv
```

Figure 4-58. *Creating a service definition*

4. Define the custom service fields, as shown in Figure 4-59 and save the file. Use the custom message robot_peripherals created in the previous section. The reply part is the primitive type bool. See Figure 4-59.

Figure 4-59. *Describing the custom service*

5. Open the package.xml file located in the package and add the lines highlighted in yellow, as shown in Figure 4-60. This will convert the message definition into code during compilation. This code can be used in your ROS nodes.

```
<!-- Use doc_depend for packages you need only for
building documentation: -->
  <!--   <doc_depend>doxygen</doc_depend> -->
  <buildtool_depend>catkin</buildtool_depend>
  <build_depend>roscpp</build_depend>
  <build_depend>rospy</build_depend>
  <build_depend>std_msgs</build_depend>
  <build_depend>message_generation</build_depend>
  <build_depend>actionlib_msgs</build_depend>
  <build_export_depend>roscpp</build_export_depend>
  <build_export_depend>rospy</build_export_depend>
  <build_export_depend>std_msgs</build_export_depend>
  <build_export_depend>actionlib_msgs</-
build_export_depend>
  <exec_depend>roscpp</exec_depend>
  <exec_depend>rospy</exec_depend>
  <exec_depend>std_msgs</exec_depend>
  <exec_depend>actionlib_msgs</exec_depend>
  <exec_depend>message_runtime</exec_depend>
```

Figure 4-60. *Editing the package.xml file*

6. Open the CMakeLists.txt file located in the
   package and add the message_generation and
   std_msgs lines, as shown in Figure 4-61. This helps
   generate your custom service.

```
find_package(catkin REQUIRED COMPONENTS
  roscpp
  rospy
  std_msgs
  actionlib_msgs
  message_generation
)
```

Figure 4-61. *Editing the CMakeLists.txt file*

7. In `CMakeLists.txt`, add the `message_runtime` and `std_msgs` lines, which are required during runtime. See Figure 4-62.

```
## DEPENDS: system dependencies of this project that
dependent projects also need
catkin_package(
#   INCLUDE_DIRS include
#   LIBRARIES chapter3
    CATKIN_DEPENDS roscpp rospy std_msgs actionlib_msgs
message_runtime
#   DEPENDS system_lib
)
```

Figure 4-62.   *Editing the CMakeLists.txt file*

8. Also in the `CMakeLists.txt` file, add the section shown in Figure 4-63.

```
## Generate services in the 'srv' folder
add_service_files(
  FILES
  robot_accessories.srv
)
```

Figure 4-63.   *Editing the CMakeLists.txt file*

9. You need to specify the required components (dependencies) by adding `std_msgs` in the `CMakeLists.txt` file, as shown in Figure 4-64.

```
## Generate added messages and services with any
dependencies listed here
generate_messages(
    DEPENDENCIES
    std_msgs
    actionlib_msgs
)
```

Figure 4-64. *Editing the CMakeLists.txt file*

10. Build the workspace using the catkin_make or catkin build command.

11. Check if the service has been generated. See Figure 4-65.

```
rajesh@ubuntu:~$ rossrv show robot_accessories
[chapter3/robot_accessories]:
chapter3/robot_peripherals request_msg
  bool led_front
  bool led_left
  bool led_right
  bool led_rear
  int8 music_id
  string screen_message
---
bool respone_msg
```

Figure 4-65. *Fields of the custom service*

In Figure 4-65, you can see that the custom service has two fields. They are as follows:

- request_msg: The request part is of type robot_peripherals. It is the custom message created in the previous section.

- response_msg: The reply part, which is of the bool type, indicates the success/failure of the operation.

This section explains how to create a service server and a service client using the custom service type in your ROS nodes.

Writing a Custom Service Server

```
1. #!/usr/bin/env python3
2. '''Service Server Code'''
3.
4. import rospy
5. from chapter3.srv import robot_accessories, robot_
   accessoriesResponse
6.
7. def handle_switch(req):
8.     '''Service handler'''
9.     rospy.loginfo("Received Request:")
10.    rospy.loginfo("led_front: " + str(req.request_msg.
       led_front))
11.    rospy.loginfo("led_left: " + str(req.request_msg.
       led_left))
12.    rospy.loginfo("led_right: " + str(req.request_msg.
       led_right))
13.    rospy.loginfo("led_rear: " + str(req.request_msg.
       led_rear))
14.    rospy.loginfo("music_id: " + str(req.request_msg.
       music_id))
15.    rospy.loginfo("screen_message: " + str(req.request_msg.
       screen_message)+"\n")
16.    rospy.loginfo("Operation Complete !")
17.
18.    msg = robot_accessoriesResponse()
19.    msg.respone_msg = True
20.    return msg
```

```
21.
22. if __name__ == "__main__":
23.     rospy.init_node('robot_accessories_server')
24.     s = rospy.Service('/robot_accessories_controller',
        robot_accessories, handle_switch)
25.     print("Ready to control robot accessories ...")
26.     rospy.spin()
27.
```

This code waits for the client's input in the custom service format. Line 5 imports the custom service messages. Lines 10-15 display the fields of the request message. Line 18 creates a variable of type custom service response. Line 19 fills the field of the message with a True value, indicating that the operation was successful.

Writing a Custom Service Client

```
1. #!/usr/bin/env python3
2. '''Service Client Code'''
3.
4. import rospy
5. from chapter3.srv import robot_accessories, robot_
   accessoriesRequest, robot_accessoriesResponse
6.
7. if __name__ == '__main__':
8.     rospy.init_node("robot_accessories_client")
9.     rospy.wait_for_service("/robot_accessories_controller")
10.
11.     try:
12.         response = robot_accessoriesResponse()
13.         robot_accessories_service = rospy.ServiceProxy("/
            robot_accessories_controller", robot_accessories)
```

```
14.         led_front = input("Enter ON/OFF for led_front: ")
15.         led_left = input("Enter ON/OFF for led_left: ")
16.         led_right = input("Enter ON/OFF for led_right: ")
17.         led_rear = input("Enter ON/OFF for led_rear: ")
18.         music_id = input("Enter a number for music_id: ")
19.         screen_message = input("Enter a text for screen_
            message: ")
20.
21.         req = robot_accessoriesRequest()
22.
23.         req.request_msg.led_front = False
24.         req.request_msg.led_left = False
25.         req.request_msg.led_right = False
26.         req.request_msg.led_rear = False
27.
28.         if(led_front == "ON"):
29.             req.request_msg.led_front = True
30.         if(led_left == "ON"):
31.             req.request_msg.led_left = True
32.         if(led_right == "ON"):
33.             req.request_msg.led_right = True
34.         if(led_rear == "ON"):
35.             req.request_msg.led_rear = True
36.
37.         req.request_msg.music_id = int(music_id)
38.         req.request_msg.screen_message = screen_message
39.
40.         response = robot_accessories_service(req)
            #service call
41.         rospy.loginfo("Response from Server: " +
            str(response.respone_msg))
```

```
42.        except rospy.ServiceException as e:
43.            rospy.logwarn("Service failed: " + str(e))
44.            rospy.loginfo("Response from Server: Exception !")
45.
```

This code collects input from the user via the keyboard and sends it to the server, triggering some functionalities of the robot. Line 5 imports the custom service messages. Line 21 creates a variable of type custom service request. Lines 23-38 fill the fields of the request message according to the user input. On Line 40, the client requests the server to perform operations and waits for the response.

Figures 4-66 through 4-68 show the output.

Figure 4-66. *Commands to run the custom service server and client*

Figure 4-67. *Client prompting inputs from the user via the keyboard*

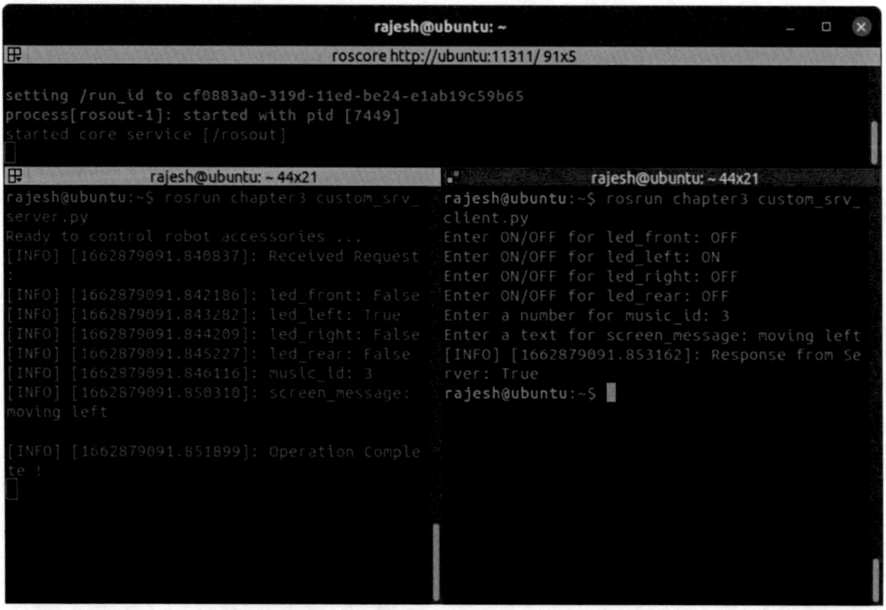

Figure 4-68. *Server performing operations and returning a result*

Create a Custom Action Definition

An *action definition* is a plain text file that contains three parts—a goal, results, and feedback. Each part is separated by - - -. These parts can contain one or more fields and each field has a type and name. The types that can be used in the action definition are as follows:

- Primitive types

    - bool

    - int8

    - int16

    - int32

    - int64

    - uint8

- uint16

- uint32

- uint64

- float32

- float64

- string

- time

- duration

- Arrays (i.e., list of primitive types, such as int8[ ], bool[ ], etc.)

- Other messages (for example, messages provided by the std_msgs package, custom messages, etc.)

To learn more about primitive datatypes in ROS, see http://wiki. ros.org/msg.

An example of a predefined action in ROS is the Averaging action, which is provided by the actionlib_tutorials library. You can see the contents of this action definition in Figure 4-69.

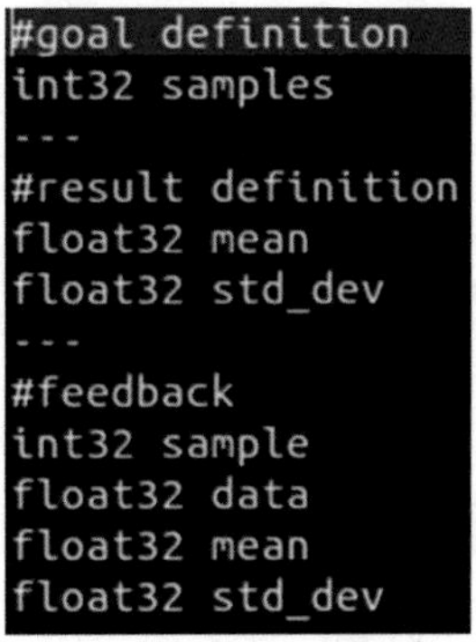

```
#goal definition
int32 samples
- - -
#result definition
float32 mean
float32 std_dev
- - -
#feedback
int32 sample
float32 data
float32 mean
float32 std_dev
```

Figure 4-69. *Definition of the Averaging message*

The `Averaging` action message has three parts, separated by the `---` symbol. This symbol indicates the separation between the goal, result, and feedback parts. The first part is the goal, the second part is the result, and the third part is the feedback. The field types used here are `int32` and `float32`, which are primitive message types.

Procedure

Create a custom action to move the robot and get a response. Follow these steps to do so:

12. Navigate to the package, as shown in Figure 4-70.

```
rajesh@ubuntu:~$ roscd chapter3
rajesh@ubuntu:~/ROS_Book_WS/src/chapter3$
```

***Figure 4-70.** Navigating to the package*

13. Create a folder called `action`, as shown in Figure 4-71.

```
rajesh@ubuntu:~/ROS_Book_WS/src/chapter3$ mkdir action
```

***Figure 4-71.** Creating the action folder*

14. Within the `action` folder, create a new file called `<action_name>.action`, where `< action_name>` is the name of the action you intend to create. This example creates a custom action called `MoveRobot`, as shown in Figure 4-72.

```
rajesh@ubuntu:~/ROS_Book_WS/src/chapter3/action$ gedit MoveRobot.action
```

***Figure 4-72.** Creating the action definition*

15. Define the custom action fields as shown in
    Figure 4-73 and save the file. You now have goal,
    the result, and feedback parts, which are of string
    list, string, and string list types, respectively.
    See Figure 4-73.

Figure 4-73. *Describing the custom action*

16. Open the package.xml file located in the package
    and add the lines highlighted in yellow in
    Figure 4-74. This will convert the action definition
    into code during compilation. This code can be used
    in your ROS nodes.

```
<!-- Use doc_depend for packages you need only for
building documentation: -->
<!--     <doc_depend>doxygen</doc_depend> -->
<buildtool_depend>catkin</buildtool_depend>
<build_depend>roscpp</build_depend>
<build_depend>rospy</build_depend>
<build_depend>std_msgs</build_depend>
<build_depend>message_generation</build_depend>
<build_depend>actionlib_msgs</build_depend>
<build_export_depend>roscpp</build_export_depend>
<build_export_depend>rospy</build_export_depend>
<build_export_depend>std_msgs</build_export_depend>
<build_export_depend>actionlib_msgs</build_export_depend>
<exec_depend>roscpp</exec_depend>
<exec_depend>rospy</exec_depend>
<exec_depend>std_msgs</exec_depend>
<exec_depend>actionlib_msgs</exec_depend>
<exec_depend>message_runtime</exec_depend>
```

Figure 4-74. *Editing the package.xml file*

17. Open the CMakeLists.txt file located in the package and add the actionlib_msgs, std_msgs, and message_generation lines, as shown in Figure 4-75. This helps generate the custom action.

```
find_package(catkin REQUIRED COMPONENTS
  roscpp
  rospy
  std_msgs
  actionlib_msgs
  message_generation
)
```

Figure 4-75. *Editing the CMakeLists.txt file*

18.   In the CMakeLists.txt file, add the std_msgs,
       actionlib_msgs, and message_runtime lines, which
       are the required components (dependencies). See
       Figure 4-76.

```
## DEPENDS: system dependencies of this project that
dependent projects also need
catkin_package(
#    INCLUDE_DIRS include
#    LIBRARIES chapter3
    CATKIN_DEPENDS roscpp rospy std_msgs actionlib_msgs
message_runtime
#    DEPENDS system_lib
)
```

Figure 4-76. *Editing the CMakeLists.txt file*

19.   Also, within CMakeLists.txt file, add the section
       shown in Figure 4-77.

```
## Generate actions in the 'action' folder
 add_action_files(
     FILES
     MoveRobot.action
 )
```

Figure 4-77. *Editing the CMakeLists.txt file*

20. You need to specify dependencies by adding std_
    msgs and actionlib_msgs to the CMakeLists.txt
    file, as shown in Figure 4-78.

```
## Generate added messages and services with any
dependencies listed here
 generate_messages(
    DEPENDENCIES
    std_msgs
    actionlib_msgs
 )
```

Figure 4-78.   *Editing the CMakeLists.txt file*

21. Build the workspace using the catkin_make or
    catkin build command.

22. Check if the action is generated. See Figure 4-79.

```
rajesh@ubuntu:~/ROS_Book_WS$ rosmsg show MoveRobotAction
[chapter3/MoveRobotAction]:
chapter3/MoveRobotActionGoal action_goal
  std_msgs/Header header
    uint32 seq
    time stamp
    string frame_id
  actionlib_msgs/GoalID goal_id
    time stamp
    string id
  chapter3/MoveRobotGoal goal
    string[] goals
chapter3/MoveRobotActionResult action_result
  std_msgs/Header header
    uint32 seq
    time stamp
    string frame_id
  actionlib_msgs/GoalStatus status
    uint8 PENDING=0
    uint8 ACTIVE=1
    uint8 PREEMPTED=2
    uint8 SUCCEEDED=3
    uint8 ABORTED=4
    uint8 REJECTED=5
    uint8 PREEMPTING=6
    uint8 RECALLING=7
    uint8 RECALLED=8
    uint8 LOST=9
    actionlib_msgs/GoalID goal_id
      time stamp
      string id
    uint8 status
    string text
  chapter3/MoveRobotResult result
    string result
chapter3/MoveRobotActionFeedback action_feedback
  std_msgs/Header header
    uint32 seq
    time stamp
    string frame_id
  actionlib_msgs/GoalStatus status
    uint8 PENDING=0
    uint8 ACTIVE=1
    uint8 PREEMPTED=2
    uint8 SUCCEEDED=3
    uint8 ABORTED=4
    uint8 REJECTED=5
    uint8 PREEMPTING=6
    uint8 RECALLING=7
    uint8 RECALLED=8
    uint8 LOST=9
    actionlib_msgs/GoalID goal_id
      time stamp
      string id
    uint8 status
    string text
  chapter3/MoveRobotFeedback feedback
    string[] goals_reached_till_now
```

Figure 4-79. *Fields of the custom action message*

In Figure 4-79, you can see that the custom service has three fields. They are as follows:

- action_goal: The goal part is of type MoveRobotActionGoal. It indicates the operations requested by the client.

- action_result: The reply part is of the MoveRobotActionResult type. It indicates the result of the operation and is sent from the server to the client after performing the operation. It is only sent once.

- action_feedback: Intermediate messages are sent from the server to the client to indicate the progress of the operations. It is of type MoveRobotActionFeedback.

Note   These three action message fields are generated automatically (during the build operation) from the custom action definition MoveRobot.action, as shown in Figure 4-73.

This section explains how to create an action server and action client using the custom action type in your ROS nodes.

Creating a Custom Action Server

```
1. #!/usr/bin/env python3
2. '''Action Server Code'''
3.
4. import time
5. import actionlib
6. import rospy
7. from chapter3.msg import MoveRobotAction
8. from chapter3.msg import MoveRobotFeedback
```

```
9. from chapter3.msg import MoveRobotResult
10.
11. ACTION_SERVER = None
12.
13. def move_robot_callback(received_goals):
14.     ''' Callback '''
15.     global ACTION_SERVER
16.     reached_goals = []
17.     delay = 3 #wait time in seconds
18.     rate = rospy.Rate(1)
19.     rospy.loginfo("Goals have been received")
20.     rospy.loginfo(received_goals)
21.     success = False
22.     preempted = False
23.     i = 0
24.
25.     while not rospy.is_shutdown():
26.         #Taking each goal one by one and moving the robot
                towards it
27.         current_goal = received_goals.goals[i]
28.         rospy.loginfo("Moving to goal " + str(current_goal)
                + " ...")
29.         time.sleep(delay) #Dummy delay
30.         rospy.loginfo("Reached goal " + str(current_goal))
31.         reached_goals.append(current_goal)
32.         i = i+1
33.         if ACTION_SERVER.is_preempt_requested():
34.             preempted = True
35.             break
36.         if reached_goals == received_goals.goals:
37.             success = True
```

```
38.                break
39.           feedback_msg = MoveRobotFeedback()
40.           feedback_msg.goals_reached_till_now = reached_goals
41.           ACTION_SERVER.publish_feedback(feedback_msg)
42.           rate.sleep()
43.
44.      result = MoveRobotResult()
45.      if success:
46.          result.result = "All goals visited !"
47.      else:
48.          result.result = "All goals not visited ..."
49.      rospy.loginfo("sent goal result to client")
50.
51.      if preempted:
52.          rospy.loginfo("Preempted")
53.          ACTION_SERVER.set_preempted(result)
54.      elif success:
55.          rospy.loginfo("Success")
56.          ACTION_SERVER.set_succeeded(result)
57.      else:
58.          rospy.loginfo("Failure")
59.          ACTION_SERVER.set_aborted(result)
60.
61. def main():
62.      '''Main'''
63.      global ACTION_SERVER
64.      ACTION_SERVER = actionlib.SimpleActionServer('/
         move_robot', MoveRobotAction, execute_cb=move_robot_
         callback, auto_start=False)
65.      ACTION_SERVER.start()
66.
```

```
67. if __name__ == '__main__':
68.     rospy.init_node('move_robot_action_server')
69.     main()
70.     rospy.spin()
71.
```

This code waits for the client's request in the custom action format. Lines 7-9 import the custom action messages. Line 39 creates a custom feedback message. Line 40 fills the feedback message field and Line 41 publishes the field. Line 44 creates a custom result message. On Lines 46 and 48, the result message field is filled accordingly. Lines 53, 56, and 59 return the result message to the client, depending on the status of the operation.

Creating a Custom Action Client

```
1. #!/usr/bin/env python3
2. '''Action Client Code'''
3.
4. import rospy
5. import actionlib
6. from chapter3.msg import MoveRobotAction
7. from chapter3.msg import MoveRobotGoal
8.
9. def done_callback(status, result):
10.     '''Completed Callback'''
11.     rospy.loginfo("Status: " + str(status))
12.     rospy.loginfo("Result: " + str(result))
13.
14. def feedback_callback(feedback):
15.     '''Status Callback'''
16.     rospy.loginfo("Feedback: " + str(feedback))
17.
```

```
18. def main(action_client):
19.      '''Main'''
20.      while(True):
21.          goals_list = input("\nEnter the goal ID/s separated
             by commas. Eg- 1,2,3 : ")
22.          if(len(goals_list)>0):
23.              goals_list = goals_list.strip().split(',')
24.              goal = MoveRobotGoal(goals=goals_list)
25.              action_client.send_goal(goal, done_cb=done_
                 callback, feedback_cb=feedback_callback)
26.              rospy.loginfo("Calling the action server")
27.              break
28.          else:
29.              continue
30.
31. if __name__ == '__main__':
32.      rospy.init_node('move_robot_action_client')
33.      ACTION_CLIENT = actionlib.SimpleActionClient('/move_
         robot', MoveRobotAction)
34.      ACTION_CLIENT.wait_for_server()
35.      rospy.loginfo("Robot waiting for goals ...")
36.      main(ACTION_CLIENT)
37.      rospy.spin()
38.
```

This code collects the list of goals from the user via the keyboard and sends it to the server. Lines 6 and 7 import the custom action messages. You have two callback functions for displaying text:

- Feedback messages from the server (feedback_callback on Line 14)

- Result message from the server (done_callback on Line 9)

Line 24 creates a custom action goal and passes the goal list as a parameter. On Line 25, the client requests the server to perform operations. The client then waits for feedback and the results.

Figures 4-80 through 4-82 show the output.

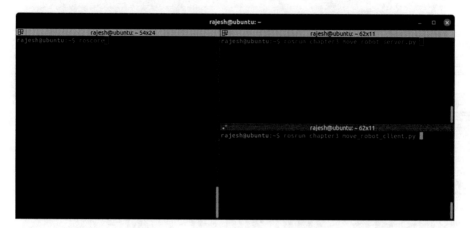

Figure 4-80. *Commands to run the ROS master, action server, and action client*

Figure 4-81. *Action client prompting the user to enter the list of goals*

Figure 4-82. *Action server performing the operations and client receiving feedback, status, and result*

Basic ROS, Catkin, and Git Commands

In a typical robotics project, you need some tools to control the robot, manage packages, build packages, maintain the version history, keep track of updates in the code, and so on.

ROS provides a set of convenient commands (or command-line tools) to control robots. The catkin tool manages and builds ROS packages. To maintain version history and keep track of updates in the workspace, you can use a popular tool called Git.

The commonly used command tools and their commands are explained in Table 4-4.

Table 4-4. *Basic Commands*

#	Command	Description	Example
1	`catkin build`	Builds a package or workspace	`catkin build`
2	`catkin config --blacklist`	Exempts the specified package from building	`catkin config --blacklist dummy_pkg`
3	`catkin config --whitelist`	Builds only the specified packages	`catkin config --whitelist dummy_pkg`
4	`catkin_create_pkg`	Creates a new ROS package	`catkin_create_pkg my_pkg rospy std_msgs`
5	`catkin_make`	Builds a ROS workspace	`catkin_make`
6	`git add -A`	Adds all the changes made in the local repository to a staging area, which is committed later	`git add -A`
7	`git branch`	Creates a new branch	`git branch new_branch`
8	`git branch -d`	Deletes a branch	`git branch -d new_branch`
9	`git branch --list`	Lists all the local branches of the repository	`git branch`

(continued)

225

Table 4-4. *(continued)*

#	Command	Description	Example
10	git checkout	Moves to another branch	git checkout new_branch
11	git checkout -b	Creates a new branch and moves to it	git checkout -b new_branch
12	git clone	Copies an existing Git repository into the computer	git clone https://github.com/logicraju/ROS_Book_WS.git
13	git commit	Saves the changes added to the staging area along with a comment	git commit -m "version1 complete"
14	git fetch	Reports changes to the remote repository	git fetch
15	git init	Sets up a new empty Git repository	git init
16	git log --oneline	Prints the updated history of the repository. Each line has a commit ID and a comment message	git log --oneline
17	git merge	Combines the update histories of the current branch and the specified branch	git merge new_branch
18	git pull	Downloads a branch in the remote repository into local	git pull origin main

(continued)

Table 4-4. (*continued*)

#	Command	Description	Example
19	git push	Uploads a branch in the local repository into remote	git push origin main
20	git remote add	Connects the local repository to the remote one	git remote add origin https://github.com/logicraju/ROS_Book_WS.git
21	git reset --hard HEAD~1	Cancels the last commit and deletes the changes	git reset --hard HEAD~1
22	git reset --soft HEAD~1	Cancels the last commit, but does not delete the changes	git reset --soft HEAD~1
23	git status	Displays the changes made in the repository since the last commit	git status
24	rosbag record -a	Records the messages under every active topic into a bag file	rosbag record -a
25	roscd	Navigates directly into the package location	roscd chapter3
26	rosclean purge	Cleans the log files generated by ROS	rosclean purge

(*continued*)

Table 4-4. (*continued*)

#	Command	Description	Example
27	roscore	Starts the master node, parameter server, and so on	roscore
28	rosdep install	Installs all the dependencies of a package or workspace	rosdep install --from-paths src --ignore-src -r -y
29	roslaunch	Runs multiple nodes using a single launch file	roslaunch turtlebot3_gazebo turtlebot3_empty_world.launch
30	rosmsg list	Displays all messages	rosmsg list
31	rosmsg package	Displays all the message types in a package	rosmsg package chapter3
32	rosmsg packages	Displays a list of packages that contain messages	rosmsg packages
33	rosmsg show	Displays the message fields of the specified message type	rosmsg show std_msgs/String
34	rosnode info	Gets information about a ROS node	rosnode info /trigger_switch_server
35	rosnode kill	Terminates a running node	rosnode kill /trigger_switch_server

(continued)

Table 4-4. (*continued*)

#	Command	Description	Example
36	rosnode list	Gets a list of all running nodes	rosnode list
37	rospack depends	Gets a list of dependencies of the package	rospack depends chapter3
38	rospack find	Gets the location of the package	rospack find chapter3
39	rosparam delete	Deletes a parameter from the parameter server	rosparam delete /robot_name
40	rosparam dump	Saves the parameters in the parameter server to a specified file	rosparam dump param_file
41	rosparam get	Gets a parameter value from the parameter server	rosparam get /robot_name
42	rosparam list	Displays a list of all parameters in the parameter server	rosparam list
43	rosparam load	Loads the parameters in the specified file into the parameter server	rosparam load param_file
44	rosparam set	Sets a parameter value into the parameter server	rosparam set /robot_name "BumbleBot"

(*continued*)

Table 4-4. (*continued*)

#	Command	Description	Example
45	rosrun	Runs an executable file (e.g., a Python script) in a package	rosrun chapter3 publisher.py
46	rosservice call	Calls an active service	rosservice call /trigger_ switch "data: false"
47	rosservice find	Gets a list of all active services of the specified type	rosservice find std_srvs/ SetBool
48	rosservice info	Gets information about an active service	rosservice info /trigger_ switch
49	rosservice list	Gets a list of active services	rosservice list
50	rossrv list	Displays all services	rossrv list
51	rossrv package	Displays all the service types in a package	rossrv package chapter3
52	rossrv packages	Displays a list of packages that contain services	rossrv packages
53	rossrv show	Displays the service definition of the specified service type	rossrv show std_srvs/SetBool

(*continued*)

Table 4-4. (*continued*)

#	Command	Description	Example
54	rostopic echo	Displays the messages published under a topic on screen	rostopic echo /pub_topic
55	rostopic find	Gets a list of all active topics of the specified type	rostopic find std_msgs/String
56	rostopic hz	Displays the rate of publishing of messages under a topic	rostopic hz /pub_topic
57	rostopic info	Gets information about a topic	rostopic info /pub_topic
58	rostopic list	Displays a list of all active topics onscreen	rostopic list
59	rostopic pub	Manually publishes messages on a topic from the terminal	rostopic pub /pub_topic std_msgs/String "data: 'hi !'"
60	rosversion -d	Gets the version of ROS installed in the system	rosversion -d
61	roswtf	Diagnoses issues in a package/metapackage/launch file	roswtf
62	rqt	Loads a graphical tool with a collection of monitoring and configuring tools	rqt

(*continued*)

Table 4-4. (*continued*)

#	Command	Description	Example
63	rqt_console	Loads a graphical tool to monitor log messages of the active nodes	rosrun rqt_console rqt_console
64	rqt_graph	Displays a graph showing the connections between various nodes. Also shows the topics between the nodes	rqt_graph
65	rqt_launchtree	Views the components of a launch file in a hierarchical manner	rosrun rqt_launchtree rqt_launchtree
66	rqt_logger_level	Sets the severity level of log messages generated by nodes	rosrun rqt_logger_level rqt_logger_level
67	rqt_plot	Displays a graph in which the values coming from a topic are plotted against time	rqt_plot
68	rqt_reconfigure	Displays all the parameters of the running nodes that can be adjusted at runtime	rosrun rqt_reconfigure rqt_reconfigure
69	rqt_top	Displays the running nodes and the resources they consume	rosrun rqt_top rqt_top

(*continued*)

Table 4-4. (*continued*)

#	Command	Description	Example
70	`rviz`	Loads a graphical tool to visualize the robot, its sensors, its algorithms, etc.	`rviz`
71	`tf view_frames`	Displays a graphical tree diagram with all the frames in the robot	`rosrun tf view_frames`
72	`tf_echo`	Displays translation and rotation between two frames	`rosrun tf tf_echo base_link lidar_frame`
73	`tf_monitor`	Displays detailed information about connections and delays between two frames	`rosrun tf tf_monitor odom base_footprint`

ROS Debugging Tools

When you create a ROS node, there is a possibility of syntax errors, logical errors, lack of proper coding standards, and so on. Also, there might be issues while integrating different nodes. There are several tools that can help you address these problems. Some of these tools are explained in the following sections.

Visual Studio Code

VSCode is a popular coding platform among ROS developers. It has several features for debugging the code, such as breakpoints, log points, data inspection, step over, step into, step out, and so on. It also has additional extensions such as pylint. Pylint points out errors, provides suggestions in Python code, and provides a coding standard. See this link to learn more about debugging in VSCode:

```
https://code.visualstudio.com/docs/editor/
debugging
```

Catkin_make/Catkin build

Both of these tools compile and build the workspace you create for your robot. During compilation, these tools provide you with information about any errors in your code.

The `Catkin` build tool has better options to debug the workspace. It selectively builds the required packages, cleans the workspace, tests packages, and so on. See the link to learn more: `https://catkin-tools.readthedocs.io/en/latest/index.html`.

RViz

RViz enables you visualize the robot's perspective of the world. It helps you view the robot model, sensor data, algorithms, and so on. It also enables you to determine if there are any issues (such as sensor data not receiving properly, algorithms not configured correctly, etc.).

Rqt_graph

Rqt_graph helps you visualize all the nodes that are currently running and determine how they communicate with each other. For instance, the `rqt_graph` of a publisher and subscriber is shown in Figure 4-83.

Figure 4-83.   *Rqt_graph of publisher and subscriber nodes*

Rosbag

This tool helps you record all the messages that are published by various nodes in the robot and then play them back later. This is useful to inspect errors in the nodes. To learn more, see `http://wiki.ros.org/rosbag/Commandline`.

Rqt_reconfigure

This tool helps you alter the parameter values to tune your algorithms. It can tune the parameter values while the nodes are running. You can also save the adjusted values in a file. A screenshot of the `rqt_reconfigure` tool is shown in Figure 4-84.

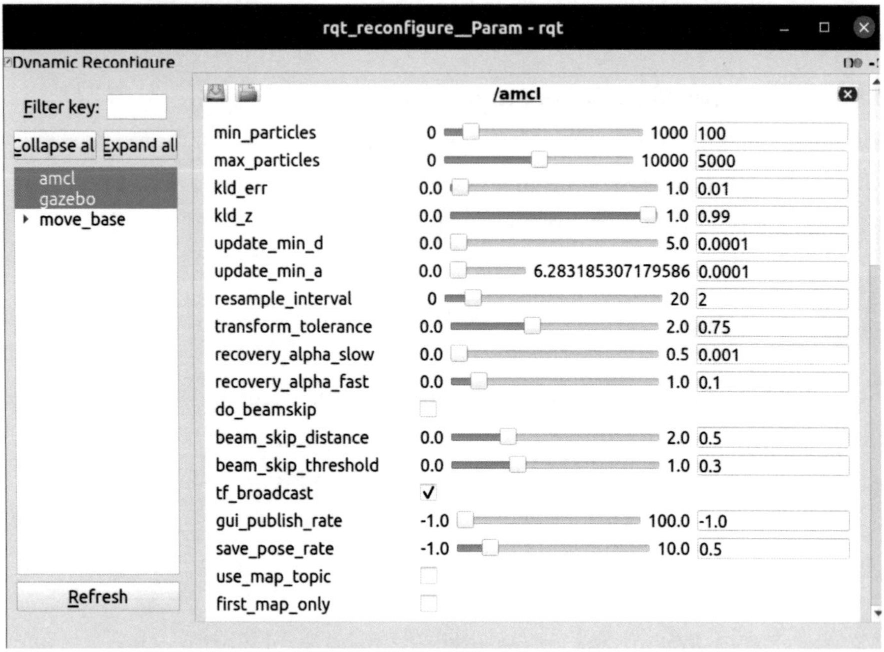

Figure 4-84. *The Rqt_reconfigure tool*

Tf View Frames

This tool gives you a graphical representation of the robot's coordinate frames. This helps you debug issues with the tf structure of the robot. A sample tf tree is shown in Figure 4-85.

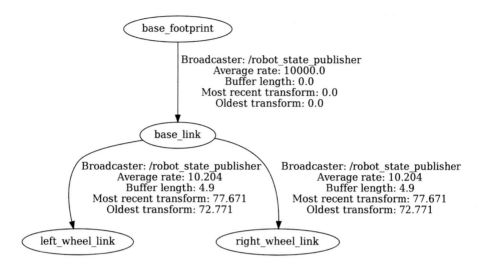

Figure 4-85.   *Example of coordinate frames in a robot*

In Figure 4-85, you can see the transformation tree of the coordinate frames of a robot. Base_footprint is the root frame, which is connected to the base_link frame. The left_wheel_link and right_wheel_link frames are the child frames of base_link.

Rqt_console

A ROS node can generate lots of log data. Logs help you understand what is happening in the node. Suppose you have a huge network of ROS nodes running and you want to keep track of the operation flow of the system. In such a scenario, looking at the log data in the terminal is hectic.

There are five types of log messages in ROS: debug, info, warn, error, and fatal. They are categorized based on their severity level. Rqt_console can filter the logs based on severity level, or via keywords to show only the relevant ones. This enables you have better insight into how the system is functioning as a whole. A screenshot is provided in Figure 4-86.

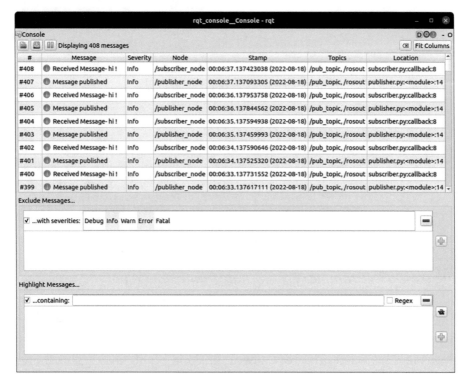

Figure 4-86. *The Rqt_console tool*

Rqt_logger_level

Usually, debug messages are not displayed in the terminal or `rqt_console`. This is because the default severity level is set to `info`. Therefore, only those messages whose severity level is greater or equal to `info` are displayed. The weight of severity levels in ascending order are as follows:

- Debug (least weight)

- Info

- Warn

- Error

- Fatal (highest weight)

The Rqt_logger_level tool allows you to specify the minimum severity level of log data. As a result, only the log data with severity levels equal to or greater that setting are considered. Figure 4-87 depicts the configuration using the rqt_logger_level tool.

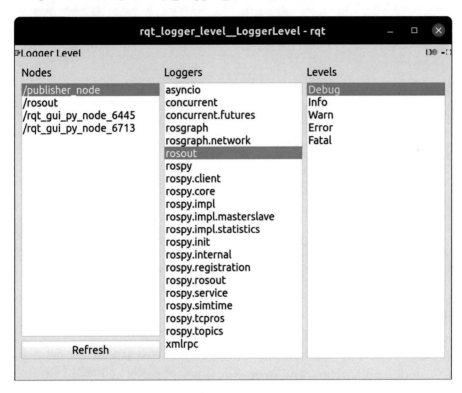

Figure 4-87. *Setting the severity level of log messages*

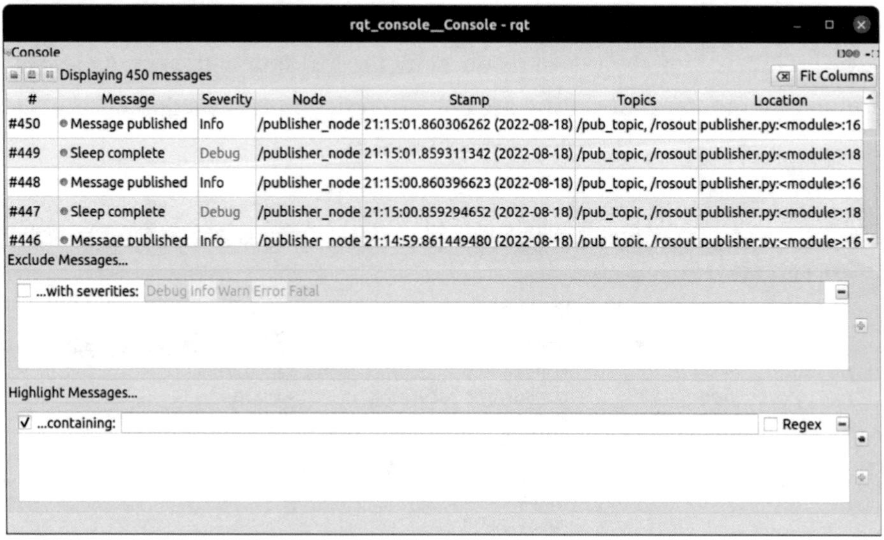

Figure 4-88. *Log messages displayed after setting the severity level*

Figure 4-88 illustrates the messages displayed after setting the log severity level. You can see that the debug messages are also displayed.

Rqt_launchtree

In a typical ROS scenario, the main launch file invokes many sub-launch files, nodes, parameters, and so on. Rqt_launchtree visualizes all these components in a tree structure. A screenshot of this tool is shown in Figure 4-89.

Note   This tool can be installed from the source into your workspace using this command:

```
git clone -b noetic https://github.com/Kuo-Feng/rqt_launchtree.git
```

You need to build your workspace using the `catkin_make` or `catkin build` command and source the workspace using the `source <path to workspace>/devel/setup.bash` command. Alternatively, you can source into your home/`<user>`/`.bashrc` file permanently, by adding the `source <path to workspace>/devel/setup.bash` line to the end of the `.bashrc` file.

Figure 4-89. *The Rqt_launchtree visualization*

Rqt

Rqt is a collection of graphical tools that visualize and tune various features of the robot. Most of these debugging tools come under this collection. Instead of launching these tools separately, you can launch `rqt` and select the required tool. It has several components, including:

241

- Action type browser

- Dynamic reconfiguration

- Launch

- Launch tree

- Node graph

- Package graph

- Process monitor

- Bag

- Console

- Logger level

- Diagnostics viewer

- Moveit! Monitor

- Robot steering

- Runtime monitor

- Service caller

- Service type browser

- Message publisher

- Message type browser

- Topic monitor

- Image view

- Navigation viewer

- Plot

- Pose view

- RViz

- Tf tree

A screenshot of this tool is shown in Figure 4-90.

Figure 4-90.   *The Rqt tool*

Rosnode/rostopic/rosservice list

These are command-line tools that do the following:

- `rosnode list`: Displays a list of currently running nodes

- `rostopic list`: Displays a list of currently running topics

- `rosservice list`: Displays a list of currently running services

Examples are provided in Figure 4-91, Figure 4-92, and Figure 4-93.

```
rajesh@ubuntu:~$ rosnode list
/publisher_node
/rosout
/subscriber_node
```

Figure 4-91. *The rosnode list command*

```
rajesh@ubuntu:~$ rostopic list
/pub_topic
/rosout
/rosout_agg
```

Figure 4-92. *The rostopic list command*

```
rajesh@ubuntu:~$ rosservice list
/rosout/get_loggers
/rosout/set_logger_level
/trigger_switch
/trigger_switch_client/get_loggers
/trigger_switch_client/set_logger_level
/trigger_switch_server/get_loggers
/trigger_switch_server/set_logger_level
```

Figure 4-93. *The rosservice list command*

Rosnode/rostopic/rosservice info

These are command-line tools that do the following:

- rosnode info: Displays the details of the currently running nodes

- rostopic info: Displays the details of the currently running topics

- rosservice info: Displays the details of the currently running services

Examples are provided in Figure 4-94, Figure 4-95, and Figure 4-96.

```
rajesh@ubuntu:~$ rosnode info /publisher_node
--------------------------------------------------------------
Node [/publisher_node]
Publications:
 * /pub_topic [std_msgs/String]
 * /rosout [rosgraph_msgs/Log]

Subscriptions: None

Services:
 * /publisher_node/get_loggers
 * /publisher_node/set_logger_level


contacting node http://ubuntu:42643/ ...
Pid: 6169
Connections:
 * topic: /pub_topic
    * to: /subscriber_node
    * direction: outbound (46407 - 127.0.0.1:46436) [11]
    * transport: TCPROS
 * topic: /rosout
    * to: /rosout
    * direction: outbound (46407 - 127.0.0.1:46434) [9]
    * transport: TCPROS
```

Figure 4-94. *The rosnode info command*

```
rajesh@ubuntu:~$ rostopic info /pub_topic
Type: std_msgs/String

Publishers:
 * /publisher_node (http://ubuntu:42643/)

Subscribers:
 * /subscriber_node (http://ubuntu:45087/)
```

Figure 4-95. *The rostopic info command*

```
rajesh@ubuntu:~$ rosservice info /trigger_switch
Node: /trigger_switch_server
URI: rosrpc://ubuntu:39861
Type: std_srvs/SetBool
Args: data
```

Figure 4-96. *The rosservice info command*

Rqt_top

`Rqt_top` displays the currently running nodes, their process identifiers, CPU use, memory use, and the number of threads. A screenshot of this tool is shown in Figure 4-97.

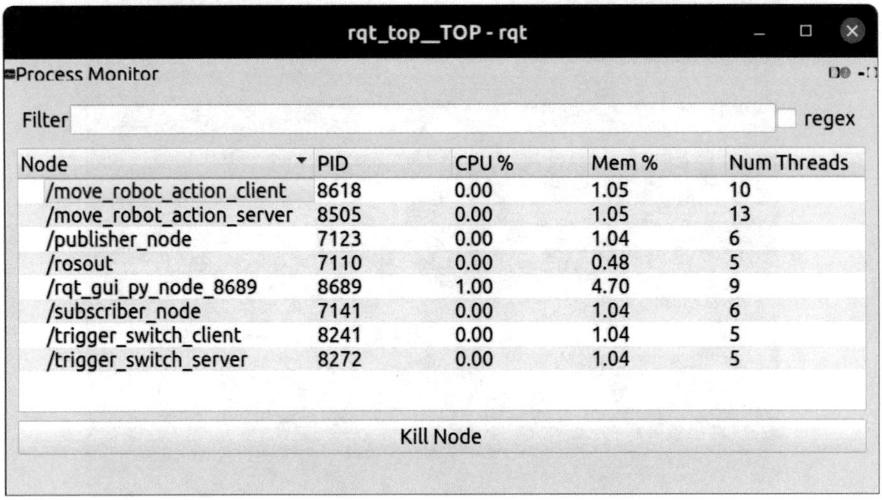
rqt_top__TOP - rqt
Process Monitor
Filter
regex
Node
PID
CPU %
Mem %
Num Threads
/move_robot_action_client 8618 0.00 1.05 10
/move_robot_action_server 8505 0.00 1.05 13
/publisher_node 7123 0.00 1.04 6
/rosout 7110 0.00 0.48 5
/rqt_gui_py_node_8689 8689 1.00 4.70 9
/subscriber_node 7141 0.00 1.04 6
/trigger_switch_client 8241 0.00 1.04 5
/trigger_switch_server 8272 0.00 1.04 5
Kill Node

Figure 4-97. *The rqt_top tool*

Roswtf

This is a command-line tool that displays any issues with the currently running nodes. The issues can include two nodes not communicating with each other, duplicate publishing of the same topic by two nodes, `tf` tree issues, and so on. A screenshot is shown in Figure 4-98.

```
rajesh@ubuntu:~$ roswtf
Loaded plugin tf.tfwtf
No package or stack in the current directory
===========================================================================
Static checks summary:

No errors or warnings
===========================================================================
Beginning tests of your ROS graph. These may take a while...
analyzing graph...
... done analyzing graph
running graph rules...
... done running graph rules
running tf checks, this will take a second...
... tf checks complete

Online checks summary:

Found 1 warning(s).
Warnings are things that may be just fine, but are sometimes at fault

WARNING No tf messages
```

Figure 4-98. *The roswtf command*

Coordinate Transformation in ROS

A robot has several interconnected links. For example, a robot can have
a base link, laser link, front left wheel link, front right wheel link, rear left
wheel link, rear right wheel link, and so on. Each link has its coordinate
frame (i.e., x, y, z axes) with origins, as shown in Figure 4-99. There could
be other frames, such as the odometry frame (initial frame used to keep
track of the robot's position), map frame (origin of the map), and so on.
These frames need to be interconnected in a specific hierarchy for the
robot's algorithms to work properly. Coordinate transformations define the
relationship between the frames. Transformation includes translation and
rotation between two frames. It converts data from one frame to another.

Example

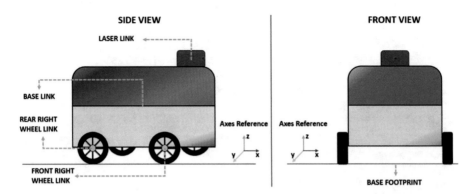

Figure 4-99. *Frames of a mobile robot*

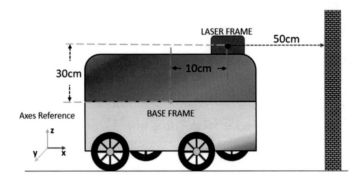

Figure 4-100. *Frame transformation example*

For example, consider the robot shown in Figure 4-100. The laser scanner mounted on top provides obstacle data (distance to obstacles with respect to the laser frame). This data has to be transformed from the laser frame to the base frame to get the distance between the robot and the obstacle. This allows the robot to avoid obstacles, plan the path, determine the robot's location on the map, and so on. Here, the laser frame is located at an offset of 30cm along the z-axis and 10cm along the x-axis from the base frame (i.e., offset is x=0.1m, y=0, z=0.3m). So, the transformation from the base frame to the laser frame can be expressed as follows:

$$Translation_{base\_to\_laser} = [0.1, 0.0, 0.3]$$

$$Rotation_{base\_to\_laser} = [0.0, 0.0, 0.0]$$

Now, the Lidar senses a wall at a distance of 50cm in front of it. So, the distance of the wall from the laser frame can be written as follows:

$$Distance\ of\ wall\ from\ laser\ frame = [0.5, 0.0, 0.0]$$

Therefore:

$$Translation_{base\_to\_wall} = Translation_{base\_to\_laser} + Distance\ of\ wall\ from\ laser\ frame$$

$$Translation_{base\_to\_wall} = [0.1, 0.0, 0.3] + [0.5, 0.0, 0.0]$$

Therefore, the transformation is as follows:

$$Translation_{base\_to\_wall} = [0.6, 0.0, 0.3]$$

$$Rotation_{base\_to\_wall} = [0.0, 0.0, 0.0]$$

In ROS, you do not have to manually calculate transformations between the frames of the robot. All the transformations between frames are described using a transformation tree (TF tree). The transformation tree is published by a ROS library called TF. From ROS Hydro version onwards, transformations are handled by a new library called TF2. The TF tree of the example robot is shown in Figure 4-101.

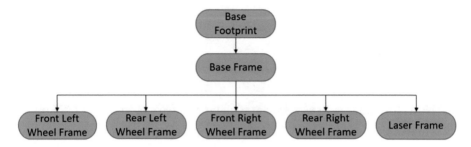

Figure 4-101. *The TF tree*

Right-Hand Rule for Coordinate Transformation

ROS uses the right-hand convention to represent the position and orientation of an object in a coordinate frame.

1. To represent the position, hold your right hand, as shown in Figure 4-102. Your index finger points to the +X axis (forward), your middle finger points to the +Y axis (leftward), and your thumb points to the +Z axis (upward).

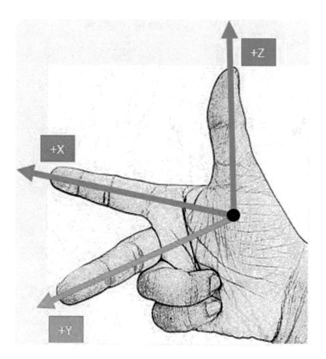

Figure 4-102. *Right-hand rule for coordinate frames*

2. To represent an orientation, hold your right hand as shown in Figure 4-103. Your thumb represents the axis of rotation and the other fingers point to the positive direction of rotation (i.e., the anticlockwise direction).

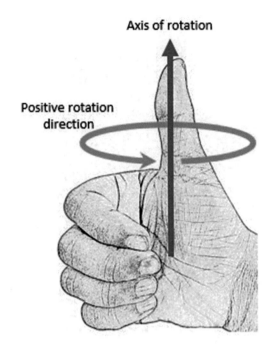

Figure 4-103. *Right-hand thumb rule*

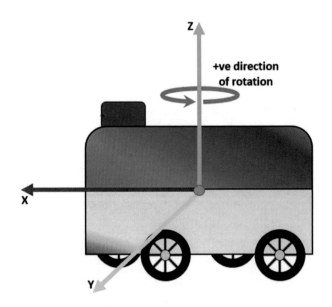

Figure 4-104. *Right-hand rule example*

For example, consider the robot shown in Figure 4-104. You want to move the robot one meter forward and take a left turn by 90 degrees.

1. To move the robot one meter forward, the robot should determine the axis to follow. The forward direction (pointed by index finger), corresponds to the positive x-axis. Therefore, the robot should move along the positive x-axis.

2. To determine the direction of rotation, hold the thumb in the direction of the +Z axis (as it is the axis of rotation of the robot to turn left/right). The robot should rotate in the direction pointed by other fingers (i.e., anticlockwise) by 90 degrees.

ROS Navigation Stack

The navigation stack in ROS is a metapackage (a collection of ROS packages) that works to bring autonomous navigation to robots. The individual components communicate with each other using messages. The main aim of the navigation stack is to get a goal pose (position and orientation) and provide velocity commands to the robot to reach the goal (see Figure 4-105).

Figure 4-105. *Navigation stack concept*

The robot continuously receives the velocity commands from the navigation stack and reaches the goal pose eventually, avoiding the obstacles along the way (see Figure 4-106).

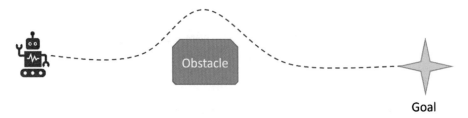

Figure 4-106. *Robot movement depiction*

There are several components required for autonomous navigation, as shown in Figure 4-107.

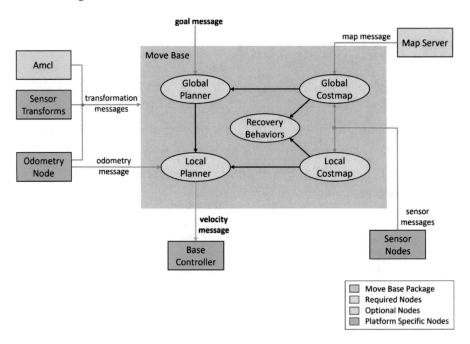

Figure 4-107. *The ROS navigation stack*

The components in the ROS navigation stack are explained next:

- The box shown in orange is the core component of the navigation stack, called the *move base*. As the name implies, the move base node deals with the movement of the robot.

- The ellipses shown in yellow are the fundamental parts of the move base package and are essential to navigation. ROS provides the global planner, global costmap, local planner, local costmap, and recovery behavior as built-in packages.

- The boxes shown in green (AMCL and map server) are optional parts.

- The boxes in blue (sensor transforms, odometry node, base controller, and sensor nodes) depend on the sensors and actuators you choose for your robot. These platform-specific nodes are sometimes provided by the manufacturer of the device. Otherwise, you have to write the nodes.

The following sections explain the components of the navigation stack in detail.

Move Base

Move base is a core component in the navigation stack. It computes the route from the initial location to the goal location and provides velocity commands to move the robot to the specified location. Move base requires sensor data, map, odometry, coordinate transformations, and localization data. There are five major components in move base. They are as follows:

- **Global planner**: This finds an optimal path from the current location to the goal location, considering static obstacles on the map. Examples of global planners provided by ROS are carrot planner, navfn, and global planner.

- **Local planner**: Divides the global path into several local goals and generates velocity commands to make the robot follow the global path. It takes into account dynamic obstacles, robot dimension, maximum velocity, maximum acceleration, and so on, to plan the local path. Examples of local planners in ROS are the base local planner, dwa planner, teb planner, and so on.

- **Global costmaps**: Represent static obstacles in the robot environment. Also inflate the obstacles on the map to prevent collision.

- **Local costmaps**: Represent the nearby obstacles of the robot using laser data. Used to avoid collisions.

- **Recovery behavior**: Set of operations done to continue navigation when the robot is stuck in the path. Default behaviors include reset, clearing rotation, aggressive reset, and aborting.

Sensor Nodes

Sensors gather information about obstacles from the environment. Sensor data is used in navigation for two purposes. One is to detect obstacles, and the other is to localize (estimate the pose of the robot on the map). To accurately localize, the sensors should have good resolution and range (such as a lidar sensor). The same sensor can also be used for obstacle avoidance. Cost-effective sensors, like infrared sensors, ultrasonic sensors, and so on, can also be used for additional obstacle avoidance by adding them to the navigation stack.

Note   You can incorporate additional obstacle avoidance by adding the sensor information to a separate layer on the map, called the range sensor layer.

ROS supports two types of sensor data formats. They are as follows:

- Laser scan: This type of data is typically obtained from a 2D lidar

- Point cloud: Obtained from a depth camera, stereo camera, and so on

Sensor Transforms

A robot has several sensors attached to it. Each sensor has its coordinate frame. The data received from the sensors needs to be transformed into the base frame (usually the robot's geometric center). This enables the robot to accurately estimate the distance between its center and the obstacle.

Odometry

Odometry refers to computing an approximate robot pose (position and orientation) with respect to the starting pose. Odometry is calculated from the data obtained from sensors, such as wheel encoders, tracking cameras, IMU, Lidar, radar, and so on. A common practice to improve accuracy in odometry is to fuse various sensor data.

Base Controller

The base controller is the part that controls the speed and direction of motors. The base controller first obtains the velocity command (including linear and angular velocities) from the local planner and splits them into individual wheel velocities. Then, it computes the required motor voltages to spin the motors at the desired speed and direction. It also obtains feedback from the encoders and corrects the errors in the motor speeds accordingly.

Map Server

The map is essential for localization and path generation. The map has information about static obstacles (i.e., those obstacles that were present at the time of creating the map; for example, walls, furniture, etc.) in the environment. The map server is the node that reads the map file and sends it to the planner in the move base node.

AMCL

Localization is the process of estimating the current robot pose on the map. AMCL stands for Adaptive Monte Carlo Localization. It is a localization technique that tracks the approximate pose of a robot on a map. AMCL takes sensor data, odometry values, and the map as inputs and estimates the location of the robot on the map. The current implementation of AMCL accepts only laser data and laser-based maps as inputs. Also, AMCL works only on 2D maps.

Summary

This chapter explained what ROS is, why you need it, and applications of ROS. You learned how to create a workspace and then learned about publishers/subscribers, services, and actions. You implemented publishers/subscribers using Python, implemented a service using Python, and implemented an action using Python. You also learned how to create custom messages, custom service definitions, and custom actions. You learned about basic ROS/Catkin/Git commands, ROS debugging tools, and coordinate transformation in ROS and ROS navigation stack.

The next chapter is about simulation and visualization using Gazebo and RViz, respectively. You also learn how to simulate and visualize various robots, including a mobile robot, a manipulator (a robotic arm), and a mobile manipulator (a robotic arm mounted to a mobile robot).

CHAPTER 5

Robot Simulation and Visualization

Outline

This chapter explains how to simulate and visualize various robots using different tools. The robots include:

- Turtlebot3: A mobile robot

- OpenMANIPULATOR-X: A robotic arm

- Turtlebot3 with OpenMANIPULATOR-X: A mobile manipulator

Simulation and Visualization

Robotic simulation is the process of approximately replicating the robot's behavior in a virtual environment. This is important, because building an actual robot takes more time, effort, and money. By using simulation, you can test and validate the robot's behavior (to an extent) before physically building it. This also minimizes damages done to hardware components

© Rajesh Subramanian 2023
R. Subramanian, *Build Autonomous Mobile Robot from Scratch using ROS*,
Maker Innovations Series, https://doi.org/10.1007/978-1-4842-9645-5_5

while testing in the real world. ROS provides a powerful robot simulation tool called Gazebo to see the actual behavior of the robot. ROS also provides a visualization tool called RViz that shows what the robot sees and "thinks."

Gazebo

Gazebo is a powerful robot simulation tool that is widely used in the robotics community. Here are some of its key features:

- Provides a highly realistic, 3D physics-based simulation environment that allows users to test and develop robotics algorithms in a safe and controlled virtual environment.

- Allows users to interact with the simulation environment, adding and removing objects (such as walls, partitions, doors, chairs, etc.), changing lighting conditions, adjusting the physics parameters, and so on.

- Supports the simulation of a variety of sensors, such as cameras, stereo cameras, depth cameras, lidars, IMUs, bumpers, and so on.

- Provides support for various drive systems, such as differential drives, skid steer drives, holonomic drives, and so on.

To start Gazebo, simply run the gazebo command in a terminal, as shown in Figure 5-1.

```
rajesh@ubuntu:~$ gazebo
```

Figure 5-1. *Gazebo launch command*

This command loads an empty Gazebo world, as shown in Figure 5-2, as you have not yet specified anything (objects, robots, etc.) to be spawned in the simulation environment.

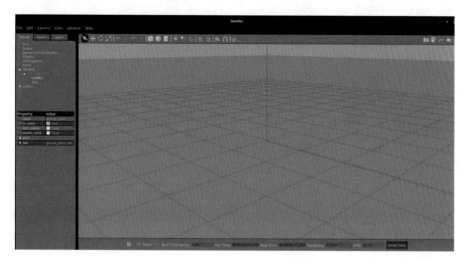

Figure 5-2. *Gazebo default world*

RViz

RViz stands for "Robot Visualization" and is a tool that visualizes several aspects of a robot, such as the sensor readings, movements, joint poses, path generated by the path planner, orientation, estimate of the current pose, and so on. RViz allows you to easily visualize and debug any issues. For example, you can visualize the path generated by a path planning algorithm in RViz. This allows you to fine-tune the algorithmic parameters to get a better path according to your requirements.

To start RViz, you need to run the ROS master using the roscore command, then use the rviz command to start RViz, as shown in Figure 5-3.

Figure 5-3. *Running RViz*

Figure 5-4. *Default RViz window*

Figure 5-4 shows the default RViz window that's launched when the rviz command is run in the terminal.

The next sections explain the simulation and visualization processes of some popular robots used in the robotics community.

Turtlebot3: Mobile Robot

Turtlebot3 is a cost-effective, compact, autonomous mobile robot made collaboratively by Robotis and Open Robotics (a non-profit organization that maintains ROS). It is mainly used in the educational and research sectors of robotics. Turtlebot3 has3 variants called Burger, Waffle, and Waffle Pi.

The next section explains how to set up the software.

Setting Turtlebot3 Up

To simulate Turtlebot3, you need to follow these steps:

1. Clone the Git repository (of this book) by running this command:

   ```
   git clone https://github.com/logicraju/ROS_
   Book_WS.git
   ```

2. Install the required dependencies using the rosdep command:

   ```
   rosdep install --from-paths src --ignore-
   src -r -y
   ```

3. Compile the entire workspace using the catkin_ make or catkin build command:

   ```
   catkin_make
   ```

4. Copy and paste the following commands to the end of the `.bashrc` file, which resides in the home folder. (Note: the bashrc file is a hidden file and can be made visible using the keyboard shortcut Ctrl+h.)

```
export TURTLEBOT3_MODEL=waffle_pi
source ~/ROS_Book_WS/devel/setup.bash
```

The first line specifies the model of Turtlebot you want to simulate. You can specify burger, waffle, or waffle_pi, as the model. The second line sets up the environment variables for the workspace that is necessary for the ROS packages and nodes to function correctly.

Simulation

After setting up the workspace, you invoke the simulation by using the following command:

```
roslaunch turtlebot3_gazebo turtlebot3_world.launch
```

This starts the simulation in Gazebo. The robot is spawned in the environment with a wooden maze, as shown in Figure 5-5.

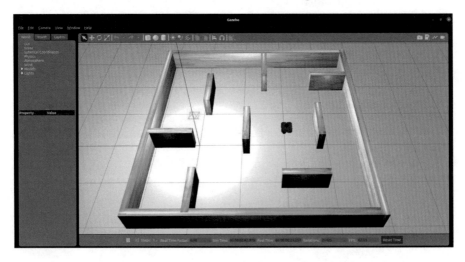

Figure 5-5. *Turtlebot3 Waffle-Pi model in a Gazebo simulation*

Teleoperation

To teleoperate the robot using the keyboard, use the following command:

```
roslaunch turtlebot3_teleop turtlebot3_teleop_key.launch
model:=waffle_pi
```

This command opens a Gazebo world with a wooden maze. Turtlebot3 is spawned as shown in Figure 5-6.

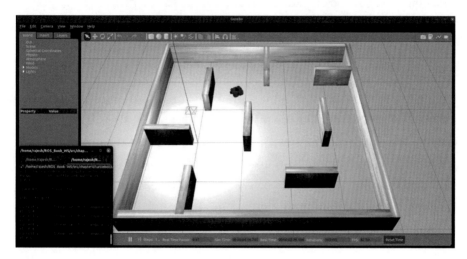

Figure 5-6. *Teleop control*

Now, you can use the keyboard keys to control the robot's movements. The keys are as follows:

- w – Move forward

- a – Turn left

- d – Turn right

- x – Move backward

- s – Stop movement

Note   The keys must be pressed within the "Teleop" window.

Mapping

To perform mapping, use the following command:

```
roslaunch turtlebot3_slam turtlebot3_slam.launch slam_
methods:=gmapping
```

This launches the required nodes for mapping along with the RViz visualizer. Gmapping is the default mapping algorithm. Other mapping algorithms (cartographer, hector slam, karto, and frontier exploration) are also available. To make the robot apply a mapping algorithm of your choice, add the cartographer, hector, karto, or frontier_exploration value for the slam_methods argument when launching. RViz shows the laser scan readings and the map created so far (see Figure 5-7). The map uses white and black colors, where white shows free space and black shows occupied space. The laser scan readings are shown in green.

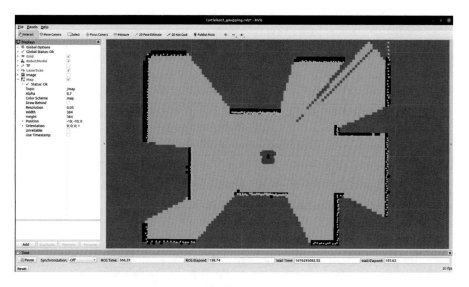

Figure 5-7. *Mapping a visualization in RViz*

To explore the environment further, move the robot using the teleoperate command. Once the environment is mapped, save the map using this command:

```
rosrun map_server map_saver -f my_map
```

Note   Replace my_map with your own map name.

Saving the map will generate two files—my_map.pgm and my_map.yaml—where the .pgm file represents the map as an image, and the .yaml file contains metadata about the map, such as resolution, origin, threshold values, and so on.

Navigation

To perform autonomous navigation, run the following command:

```
roslaunch turtlebot3_navigation turtlebot3_navigation.launch
```

To provide a goal pose for robot navigation, click the 2D Nav Goal button in RViz. Click the 2D Nav Goal button and then click and drag it to set the required goal pose on the map. The robot autonomously generates a global path, avoiding all the static obstacles on the map that block the goal. It creates a local path that follows the global path and avoids any dynamic obstacles along the way.

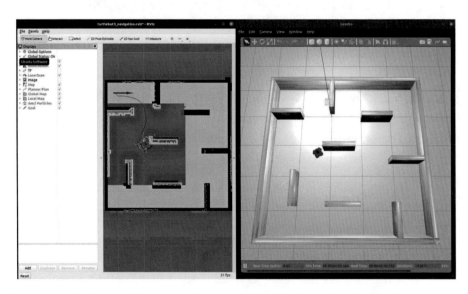

Figure 5-8. *Left: RViz visualization; Right: Gazebo simulation*

In Figure 5-8, the RViz visualizer is shown on the left. You can see the laser scan data, map, costmaps, local/global paths, goal pose, localization estimate, and so on. The Gazebo simulation is shown on the right and it depicts the actual behavior of the robot.

OpenMANIPULATOR-X: Robot Arm

OpenMANIPULATOR-X is an open-source, 5DOF robotic arm developed by Robotis and used mainly in academic and research areas. It has the following degrees of freedom:

- Base rotation along a single axis

- Shoulder rotation along a single axis

- Elbow rotation along a single axis

- Wrist rotation along a single axis

- Gripper open/close movement

You can make this robot using Dynamixel servo motors and 3D-printed parts. All the necessary CAD files are available at `www.robotis.com/service/download.php?no=767`.

Setting Up OpenMANIPULATOR-X

To simulate OpenMANIPULATOR-X, you need to follow these steps:

1. Clone the Git repository (of this book) by using this command:

   ```
   git clone https://github.com/logicraju/ROS_
   Book_WS.git
   ```

2. Install the required dependencies using the `rosdep` command:

   ```
   rosdep install --from-paths src --ignore-
   src -r -y
   ```

3. Compile the entire workspace using the `catkin_make` or `catkin build` command:

```
catkin_make
```

4. Copy and paste the following command to the end of the `.bashrc` file, which resides in the home folder. (Note: The `bashrc` file is a hidden file and can be made visible using the keyboard shortcut Ctrl+h.)

```
source ~/ROS_Book_WS/devel/setup.bash
```

This line sets up the environment variables for the workspace necessary for the ROS packages and nodes to function correctly.

Simulation

After setting up the workspace, you invoke the simulation by using the following command:

```
roslaunch open_manipulator_gazebo open_manipulator_
gazebo.launch
roslaunch open_manipulator_controller open_manipulator_
controller.launch use_platform:=false
```

This starts the simulation in Gazebo. The robot is spawned in the environment, as shown in Figure 5-9.

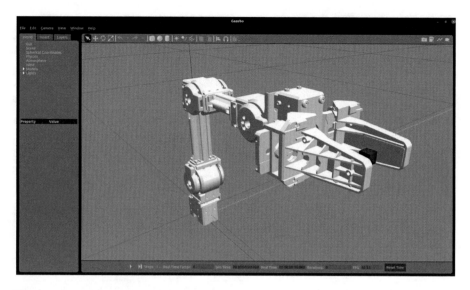

Figure 5-9. *OpenMANIPULATOR-X in the Gazebo simulator*

Click the Play button at the bottom of the Gazebo window to start the simulation.

Controlling the Arm

There are two methods available to control arm movements—one is by using a GUI application and another is by using keyboard keys.

To control the robot using a GUI program, type the following command (see Figure 5-10):

```
roslaunch open_manipulator_control_gui open_manipulator_
control_gui.launch
```

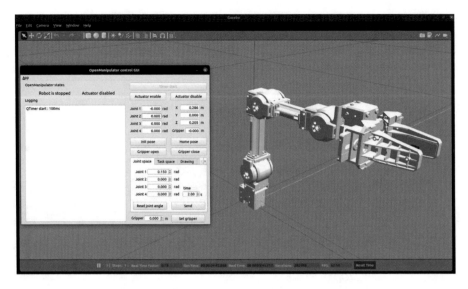

Figure 5-10. *GUI control*

Click the Timer Start button in the GUI to enable control of the joint values. Now, you can adjust the joint values in the GUI to change the joint positions. Click the Send button to transfer the values to the robot and move the robot joints accordingly.

To control the robot using the keyboard, type the following command (see Figure 5-11):

```
roslaunch open_manipulator_teleop open_manipulator_teleop_
keyboard.launch
```

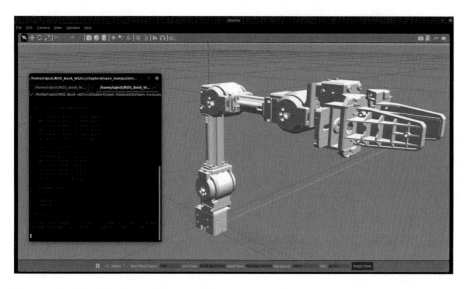

Figure 5-11. *Keyboard control*

You can now use the keyboard keys to control the robot. The keys are listed here:

- w: Increase the x-axis in the task space

- s: Decrease the x-axis in the task space

- a: Increase the y-axis in the task space

- d: Decrease the y-axis in the task space

- z: Increase the z-axis in the task space

- x: Decrease the z-axis in the task space

- y: Increase joint 1 angle

- h: Decrease joint 1 angle

- u: Increase joint 2 angle

- j: Decrease joint 2 angle

- i: Increase joint 3 angle

- k: Decrease joint 3 angle

- o: Increase joint 4 angle

- l: Decrease joint 4 angle

- g: Gripper open

- f: Gripper close

- 1: Init pose

- 2: Home pose

Note   The keys must be pressed within the "Teleop" window.

Turtlebot3 with OpenMANIPULATOR-X: Mobile Manipulator

To configure a mobile manipulator, you can mount OpenMANIPULATOR-X on top of Turtlebot3. The robot can be used to teleoperate, create maps, autonomously navigate, and pick up and place small objects at desired locations.

Setting Up

To simulate Turtlebot3 with OpenMANIPULATOR-X, you need to follow these steps:

1. Clone the Git repository (of this book) using the following command:

   ```
   git clone https://github.com/logicraju/ROS_
   Book_WS.git
   ```

2. Install the required dependencies using the rosdep command:

```
rosdep install --from-paths src --ignore-
src -r -y
```

3. Compile the entire workspace using the catkin_make or catkin build command:

```
catkin_make
```

4. Copy and paste the following commands to the end of the .bashrc file, which resides in the home folder. (Note: The bashrc file is a hidden file and can be made visible using the keyboard shortcut Ctrl+h.)

```
export TURTLEBOT3_MODEL=waffle_pi
source ~/ROS_Book_WS/devel/setup.bash
```

The first line specifies the model of the Turtlebot you want to simulate. You can specify waffle or waffle_pi as the model (the burger model cannot be used as a mobile manipulator). The second line sets up the environment variables for the workspace necessary for the ROS packages and nodes to function correctly.

Simulation

After setting up the workspace, you invoke the simulation by typing the following commands:

```
roslaunch turtlebot3_manipulation_gazebo turtlebot3_
manipulation_gazebo_maze.launch
roslaunch turtlebot3_manipulation_moveit_config move_
group.launch
```

This starts the simulation in Gazebo. The robot is spawned in the environment, as shown in Figure 5-12.

Figure 5-12. *Turtlebot3 with OpenMANIPULATOR-X in Gazebo simulation*

Click the Play button at the bottom of the Gazebo window to start the simulation.

Visualization and Motion Planning

To visualize the robot in RViz and perform motion planning of the manipulator, use the following command:

```
roslaunch turtlebot3_manipulation_moveit_config moveit_
rviz.launch
```

This launches RViz (see Figure 5-13), displaying the robot with an interactive marker attached to the end-effector of the robot arm. An *interactive marker* is a tool for interacting with 3D objects in RViz. It enables you to move and rotate robot components.

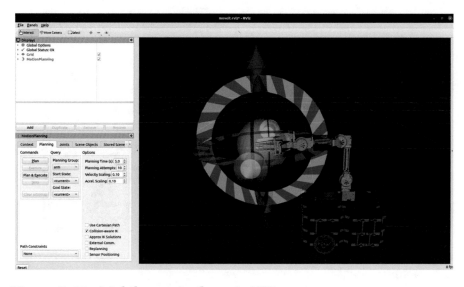

Figure 5-13. *Mobile manipulator in RViz*

To move the robot arm of the mobile manipulator to a desired pose, you can click and drag the interactive marker. Then to perform motion planning and move the arm, you need to click the Plan & Execute button on the left side of the Planning tab. This will generate the required trajectory for the manipulator and perform motion, which can be seen in RViz and Gazebo (see Figure 5-14).

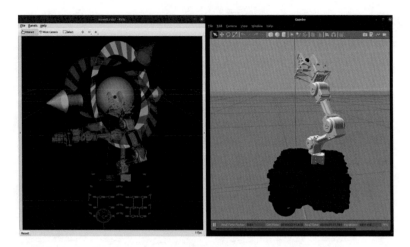

Figure 5-14. *Robot performing the motion. Left: RViz window; Right: Gazebo window*

Teleoperation

To teleoperate the robot using the keyboard, use the following command (see Figure 5-15):

```
roslaunch turtlebot3_teleop turtlebot3_teleop_key.launch
model:=waffle_pi
```

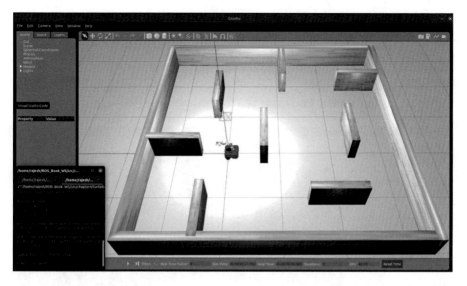

Figure 5-15. *Teleop control*

You can now use the keyboard keys to control the robot's movements. The keys are listed here:

- w – Move forward

- a – Turn left

- d – Turn right

- x – Move backward

- s – Stop movement

Note   The keys must be pressed within the "Teleop" window.

Mapping

To perform mapping, use the following command:

```
roslaunch turtlebot3_manipulation_slam slam.launch
```

281

This launches the required nodes for mapping along with the RViz visualizer. The mapping algorithm is called Gmapping. RViz shows the laser scan readings and the map created so far (see Figure 5-16). The map uses white and black colors, where white shows free space and black shows occupied space. The laser scan readings are shown in red.

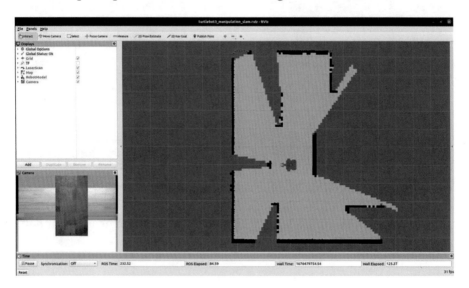

Figure 5-16. *Mapping visualization in RViz*

To explore the environment further, move the robot using the teleoperate command. Once the environment is mapped enough, save the map using this command:

```
rosrun map_server map_saver -f my_map
```

Note   Replace the my_map with your own map name.

Saving the map will generate two files called my_map.pgm and my_map.yaml, where the .pgm file represents the map as an image, and the .yaml file contains metadata about the map, such as resolution, origin, threshold values, and so on.

Navigation

To perform autonomous navigation, run the following command:

```
roslaunch turtlebot3_manipulation_navigation navigation.launch
```

To provide a goal pose for robot navigation, use the 2D Nav Goal button in RViz. Click the 2D Nav Goal button and click and drag it to set the required goal pose on the map. The robot autonomously generates a global path, avoiding all the static obstacles on the map. It also creates a local path that follows the global path and avoids any dynamic obstacles.

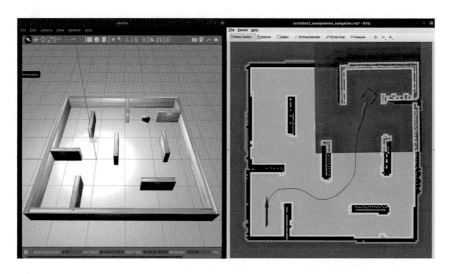

Figure 5-17. *Left: Gazebo simulation; Right: RViz visualization*

In Figure 5-17, the RViz visualizer is shown on the right, where you can see the laser scan data, map, costmaps, local/global paths, goal pose, localization estimate, and so on. The Gazebo simulation is shown on the left and it depicts the actual behavior of the robot.

Summary

This chapter explained simulation and visualization using Gazebo and RViz, respectively. It also explained simulating and visualizing various robots, including a mobile robot, a manipulator, and a mobile manipulator. In the next chapter, you look at the basics of Arduino, programming Arduino examples, interfacing Arduino with ROS, and examples of integrating Arduino with ROS.

CHAPTER 6

Arduino and ROS

Outline

This chapter covers the following topics:

- Arduino basics

- Arduino programming basics and examples

- Interfacing Arduino with ROS

- Arduino and ROS integration examples

Arduino Basics

A robot requires various electronic components such as sensors, actuators, and so on. These devices can be connected to an Arduino board, which allows you to gather inputs from sensors, control the actuators, and so on. A simple robot can work solely based on an Arduino board, whereas a complex robot can use an Arduino board as a low-level control mechanism.

Arduino is a very popular, open-source single-board microcontroller (SBM) platform. It includes of a microcontroller along with other electronic components such as input/output circuits, a clock generator,

© Rajesh Subramanian 2023
R. Subramanian, *Build Autonomous Mobile Robot from Scratch using ROS*,
Maker Innovations Series, https://doi.org/10.1007/978-1-4842-9645-5_6

RAM, ICs, and so on, to perform various operations. It's easy to learn and use, is relatively inexpensive, and is ideal for beginners and advanced users alike. Additionally, there is a large community of users who share knowledge and resources, which enables the users to find help and fix their issues quickly. A variety of electronic projects can be created using the Arduino platform, which is popular among hobbyists, educators, and professionals. Applications include robotics, home automation, traffic signaling, medical equipment and so on. See Figure 6-1.

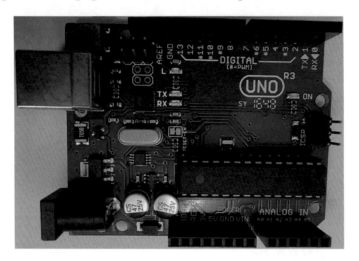

Figure 6-1. *Arduino Uno board*

Arduino Board Models

There are various models of Arduino boards, each suited to different purposes. Some of the most popular Arduino boards include:

- Arduino Uno
- Arduino Mega
- Arduino Due
- Arduino Nano

Arduino Uno (Rev3)

This board is a good choice for beginners and is suitable for small projects (see Figure 6-1). Arduino Uno's features are as follows:

- ATmega328P microcontroller
- 14 digital I/O pins (six pins for PWM)
- Six analog inputs
- One serial port (UART)
- 16MHz ceramic resonator
- USB connector
- Power jack
- ICSP header
- Reset button
- Powered using a USB (type B) cable, a DC power connector, or a battery

Arduino Mega (2560 Rev3)

Arduino Mega is the superior version of Arduino Uno. It has better specifications. The main features are as follows:

- ATmega2560 microcontroller
- 54 digital input/output pins (15 pins for PWM)
- 16 analog inputs
- Four hardware serial ports (UARTs)
- 16MHz crystal oscillator
- USB connector

- Power jack

- ICSP header

- Reset button

- Powered using a USB (type B) cable, a DC power connector, or a battery

Arduino Due

This board is the first 32-bit ARM core-based board made by Arduino. The features are as follows:

- Atmel SAM3X8E ARM Cortex-M3 processor

- 12 analog inputs

- 54 digital input/output pins (12 pins for PWM)

- Four hardware serial ports (UARTs)

- 84MHz clock

- USB OTG compatibility

- Two DAC (digital to analog) connectors

- Two TWI connectors

- Power jack

- SPI header

- JTAG header

- Reset button

- Erase button

- Powered using a micro-USB cable, a DC power connector, or a battery

> **Note**   The Arduino Due board operates at 3.3V instead of the
> usual 5V.

Arduino Nano

This is a very compact board that has approximately the same functionality
as Arduino Uno. The main characteristics are as follows:

- ATmega328 microcontroller

- Eight analog inputs

- 22 digital input/output pins (six pins for PWM)

- One hardware serial port (UART)

- 16MHz clock

- Power jack

- SPI header

- JTAG header

- Reset button

- Erase button

- Powered using a mini-B USB cable or a battery

Arduino Programming Basics

To program an Arduino, you need to write code in the Arduino
programming language (which is based on C++) and upload it into the
board via a USB port. The fundamental procedures for programming an
Arduino board are as follows:

1. Download and install Arduino IDE on your computer. Arduino IDE is the software used to write code and upload it to the microcontroller. It can be downloaded for free from the Arduino website (see www.arduino.cc/en/software). In Figure 6-2, you can see the Arduino IDE with template code.

Figure 6-2. *Arduino IDE*

2. Connect the Arduino board to your computer using a USB cable (of the appropriate type depending on your Arduino board).

3. Select the board type (depending on the Arduino
    board used) and serial port number (according to
    the USB port connected) in the Arduino IDE. In
    Figure 6-3, the board type is Arduino Uno and the
    port is COM3.

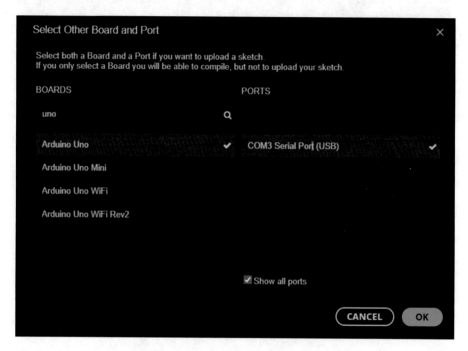

Figure 6-3. Selecting the board type and port number

4. Create the code using the C/C++ based Arduino
    programming language. Many prebuilt utilities and
    libraries that come with the Arduino IDE make it
    simple to write code. Sample code to blink the built-
    in LED of Arduino is shown in Figure 6-4.

Figure 6-4. *Sample code to blink the built-in LED of the Arduino board*

5. To compile the code, click the Verify button in the
    IDE. This will show any syntax errors. See Figure 6-5.

Verify
Button

Figure 6-5. *Use the Verify button to compile the Arduino code*

6. After the compilation is complete and has no errors,
    you need to upload the code to the Arduino board.
    This is done using the Upload button in the IDE. See
    Figure 6-6.

Upload
Button

Figure 6-6. *The Upload button transfers code to the board*

7. Test the code after connecting the required
    components to the board and power it on.

To work with Arduino, you can either use the real hardware or a
simulated version. This section of the book uses a free online electronic
simulator tool called Tinkercad Circuits. It allows users to simulate and
construct circuits using various electronic components (including an
Arduino board). It is made by the well-known software provider Autodesk,
which creates 3D design, engineering, and entertainment software.
Tinkercad Circuits enables users to virtually design and test electronic
circuits without having to buy components. To build the circuits, users
can drag and drop components like resistors, capacitors, transistors,

LEDs, and so on using a user-friendly interface in the Tinkercad software. The circuits can be tested using the simulation feature provided by the platform. Simulation allows the users to minimize damages (to electronic components), time, and money while creating a physical circuit.

Arduino Examples

This section goes through some example projects using the Arduino board.

LED Blink

In this example, an LED is made to blink five times with a delay of half a second between each flash. The circuit diagram is shown in Figure 6-7.

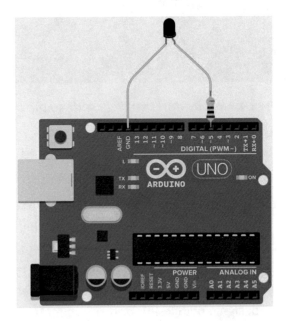

Figure 6-7. *Circuit to blink an LED using Arduino*

The anode (+ve pin) of the LED is connected to the digital pin 5 of the Arduino via a 100-ohm resistor (which controls excessive current to prevent the LED from burning out). The cathode (-ve pin) of the LED is connected to the ground pin.

Here is the code:

```
1. // LED
2. #define pin_number 5
3. int count = 0, no_of_times = 5, wait_time = 500;
4.
5. void setup()
6. {
7.   pinMode(pin_number, OUTPUT);
8. }
9.
10. void loop()
11. {
12.   if (count < no_of_times)
13.   {
14.     digitalWrite(pin_number, HIGH);
15.     delay(wait_time);
16.     digitalWrite(pin_number, LOW);
17.     delay(wait_time);
18.     count++;
19.   }
20. }
```

This Arduino program works as follows:

- Line 2 defines a macro called pin_number. The macro will be substituted with the value 5 for all occurrences during compilation. This macro indicates the pin number of Arduino to control the LED.

- Line 3 declares and initializes some variables:

  - count keeps track of the number of times the LED has blinked.

  - no_of_times defines how many times the LED needs to blink.

  - wait_time represents the delay between each LED blink (in milliseconds).

- The setup() method initializes the program by executing the statements within that block just once, at the beginning.

- The pinMode() method in the setup function designates pin 5 of Arduino as the output pin.

- The loop() method executes the statements within it as long as the Arduino is powered ON.

- If the value of count is less than no_of_times, the digitalWrite() method with the parameter HIGH sets the pin to a high voltage level and turns the LED on.

- The delay() method waits 500 milliseconds before executing the next line.

- The digitalWrite() method with the parameter LOW sets the pin to a low voltage level and turns the LED off.

- The count variable is incremented each time the LED blinks.

- The loop exits when the count variable is equal to or greater than the value of no_of_times.

When simulating the output, you can see that the LED turns on and off five times, with a delay of 500 milliseconds between each glow. Figure 6-8 shows the LED turning on and off.

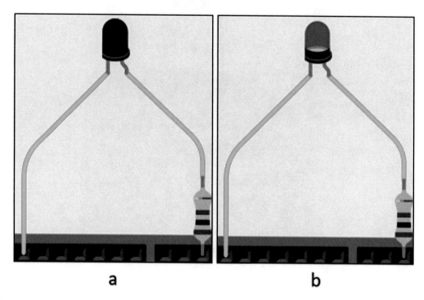

a   b

Figure 6-8.   *LED turning off (a) and on (b)*

Buzzer

In this example, a buzzer generates sounds of various frequencies five times with a delay of 100 milliseconds between each loop. The circuit diagram is shown in Figure 6-9.

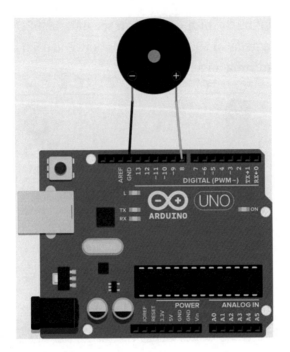

Figure 6-9. *Circuit to generate sounds using a buzzer and Arduino*

Here is the code:

```
1. // Buzzer
2. #define pin_number 8
3. int count = 0, no_of_times = 5, note_duration = 0, delay_
   time = 200, freq_start = 100, freq_end = 300;
4.
5. void setup()
6. {
7.   pinMode(pin_number, OUTPUT); // declare as output pin
8. }
9.
10. void loop()
11. {
```

```
12.    if(count<no_of_times)
13.    {
14.      for(int note=freq_start; note<freq_end; note+=25)
15.      {
16.        tone(pin_number, note, note_duration);
17.        delay(delay_time);
18.        noTone(pin_number);
19.      }
20.      count++;
21.    }
22. }
```

This Arduino program works as follows:

- Line 2 defines a macro named pin_number. This will be substituted with the value 8 for all occurrences during compilation. This macro indicates the pin number of the buzzer in Arduino.

- Line 3 declares and initializes some variables:

  - count keeps track of the number of times the buzzer sound sequence has been played.

  - no_of_times defines how many times the buzzer sound loop needs to be played.

  - note_duration represents the duration the tone should be played. (Note: Zero indicates a continuous tone.)

  - delay_time denotes the time interval between playing each frequency in milliseconds.

  - freq_start denotes the starting sound frequency (in Hertz).

- freq_end denotes the ending sound frequency (in Hertz).

- The setup() method initializes the program by executing the statements within it just once, at the beginning.

- The pinMode() method in the setup function designates pin 8 of Arduino as the output pin.

- The loop() method executes the statements within it as long as the Arduino is powered ON.

- If the value of count is less than no_of_times, the tone() method with the parameters pin_number, note, and note_duration sets the buzzer frequency accordingly.

- The delay() method waits 200 milliseconds before executing the next line.

- The noTone() method with the parameter pin_number stops

When simulating the output, you can see that the buzzer generates the sounds of the specified frequencies in a loop five times, with a delay of 200 milliseconds between each note. Figure 6-10 shows the buzzer ceasing to generate sound and then generating sound.

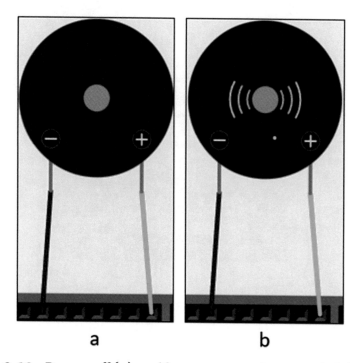

a   b

Figure 6-10. *Buzzer off (a) and buzzer generating sounds (b)*

Switch

In this example, a switch input is read and an LED is turned on/off accordingly. If the switch is pressed, the LED is turned on and vice versa. The circuit diagram is shown in Figure 6-11.

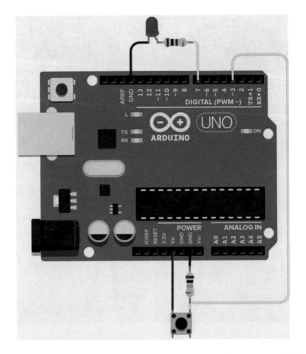

Figure 6-11. *Circuit to read switch status and turn on/off LED accordingly*

Here is the code:

```
1. // Switch
2. #define led_pin 7 // choose the pin for the LED
3. #define switch_pin 3 // choose the input pin (for a
   pushbutton)
4. int val = 0; // variable for reading the pin status
5.
6. void setup()
7. {
8.   pinMode(led_pin, OUTPUT); // declare as output pin
9.   pinMode(switch_pin, INPUT); // declare as input pin
10. }
```

```
11.
12. void loop()
13. {
14.    val = digitalRead(switch_pin); // read input value
15.    if (val == HIGH)
16.    { // check if the input is HIGH (button released)
17.      digitalWrite(led_pin, HIGH); // turn LED ON
18.    } else {
19.      digitalWrite(led_pin, LOW); // turn LED OFF } }
20.    }
21. }
```

This Arduino program works as follows:

- Lines 2 and 3 of the program define two macros named led_pin and switch_pin. These macros are substituted with the values 7 and 3, respectively, for all occurrences during compilation. These macros indicate the pin numbers of Arduino you use for the LED and switch.

- Line 4 declares and initializes the integer variable val to 0. This variable reads the status of the switch (i.e., on/off).

- The setup() method initializes the program by executing the statements within it just once, at the beginning.

- The pinMode() method in the setup function designates pin 7 of Arduino as the output pin and pin 3 of Arduino as the input pin.

- The digitalRead() method with the parameter switch_pin reads the status of the switch connected to pin 3.

- The loop() method executes the statements within it as long as the Arduino is powered ON.

- If the value of the variable val is equal to HIGH (i.e., 1), the digitalWrite() method with the parameter HIGH sets the pin to a high voltage level and turns on the LED that is attached to pin 7.

- Otherwise, the digitalWrite() method with the parameter LOW sets the pin to a low voltage level and turns off the LED.

When simulating the output, you can see that when the switch is not pressed, the LED does not glow. Whereas, when the switch is pressed, the LED glows. See Figure 6-12.

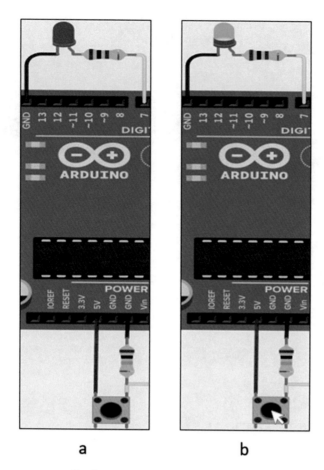

a   b

Figure 6-12. *LED off when the switch is off (a) and LED on when the switch is pressed on (b)*

LCD Display

In this example, an LCD with 2 rows and 16 columns displays some characters. The circuit diagram is shown in Figure 6-13.

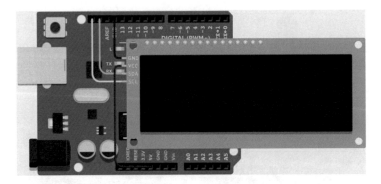

Figure 6-13. *Circuit to display characters in a 16X2 LCD*

Here is the code:

```
1. // LCD
2. #include <Adafruit_LiquidCrystal.h>
3. Adafruit_LiquidCrystal lcd(0);
4.
5. void setup()
6. {
7.    lcd.begin(16, 2);
8.    lcd.setBacklight(1);
9.    lcd.print("Hello World");
10. }
11.
12. void loop()
13. {
14.    lcd.setCursor(0, 1);
15.    lcd.print("*-*-*-*-*-*");
16.    delay(500); // Wait for 500 millisecond(s)
17.    lcd.setCursor(0, 1);
18.    lcd.print("-*-*-*-*-*-");
19.    delay(500); // Wait for 500 millisecond(s)
20. }
```

This Arduino program works as follows:

- Line 2 imports the `Adafruit_LiquidCrystal` library, which has a collection of methods to control and communicate with the LCD display.

- Line 3 creates an object named `lcd`. This object is of the class `Adafruit_LiquidCrystal` with the argument 0.

- The `setup()` method initializes the program by executing the statements within it just once, at the beginning.

- The `lcd.begin()` method initializes the LCD as 16 columns and 2 rows.

- The `lcd.setBacklight()` method sets the backlight intensity of the display.

- The `lcd.print()` method with the parameter "Hello World" displays the specified text on the LCD.

- The `loop()` method executes the statements within it as long as the Arduino is powered ON.

- The `lcd.setCursor()` method sets the cursor to the desired position.

- Line 14 sets the cursor to the second row of the LCD.

- Line 15 prints a sequence of asterisk and hyphen symbols.

- Line 16 creates a delay of 500 milliseconds.

- Line 17 sets the cursor to the second row of the LCD.

- Line 18 prints another sequence of asterisk and hyphen symbols, overwriting the previous sequence.

- Line 19 creates a delay time of 500 milliseconds.

When simulating the output, the string "Hello World" is displayed on the LCD screen. This is followed by displaying the "*-*-*-*-*-*" string in the next row for 500 milliseconds. Then, another string "-*-*-*-*-*-" is displayed (overwriting the previous sequence followed by a delay time of 500 milliseconds). This creates an animation effect in the second row of the LCD. See Figure 6-14.

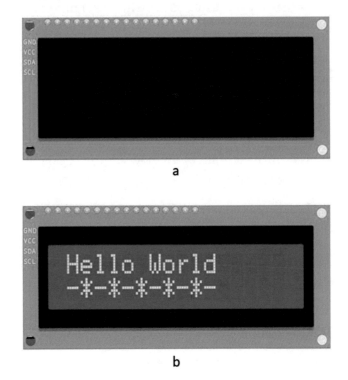

a

b

Figure 6-14. *LCD off (a) and text displayed in LCD (b)*

Interfacing Arduino with ROS

To interface Arduino with ROS-based robots, a software package called rosserial is used. It allows you to control the hardware components that are attached to the microcontroller. For this, rosserial uses serial communication between ROS nodes and the microcontroller. Using rosserial, programmers can create ROS nodes that talk to a microcontroller. This allows you to construct sophisticated robotic systems that integrate hardware and software elements.

Note   In this section, a real Arduino board is used instead of the Tinkercad Circuits simulator.

Installation

To integrate ROS with Arduino, you need to install the rosserial package and Arduino IDE.

Rosserial Installation

To install the rosserial package, use the following command:

```
sudo apt install ros-noetic-rosserial*
```

This will install all the rosserial packages for ROS Noetic version.

Arduino IDE Installation and Configuration

1. First, download Arduino IDE by visiting one of the following links:

   `https://downloads.arduino.cc/arduino-1.8.`
   `19-linux64.tar.xz` (for the 64-bit version)
   `https://downloads.arduino.cc/arduino-1.8.`
   `19-linux32.tar.xz` (for the 32-bit version)

Note   Although Arduino IDE version 2.0.4 is available, Arduino IDE version 1.8.19 is recommended to be installed in Ubuntu. This is because some bugs exist in the `roslib` library at the time of writing this book.

---

2. Extract the contents of the file into a folder.

3. Type the following commands:

   1. `cd <extracted_folder>`
   2. `sudo sh install.sh`

Now, you need to import the `rosserial` library into Arduino IDE. This can be done by navigating into the sketchbook location (in this case, `~/Arduino/libraries`) and running the Python script called `make_libraries.py` as follows:

1. `cd ~/Arduino/libraries`
2. `rosrun rosserial_arduino make_libraries.py .`

Next, ensure that the path for the sketchbook is configured correctly in the Arduino IDE, as shown in Figure 6-15.

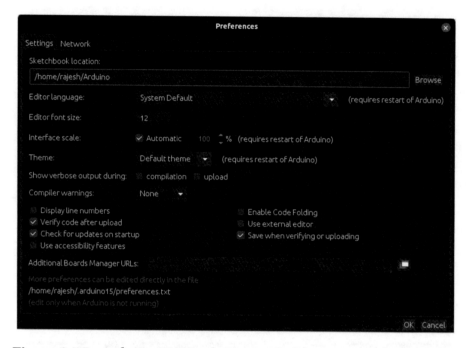

Figure 6-15. *Arduino IDE Preferences window*

Port Configuration

Next, you need to grant permission to access serial ports. This can be done using the following command:

```
sudo gpasswd --add "username" dialout
```

Note   Replace username with your actual username.

Defining Rules

To permanently set the label and the required read/write/execute permissions for the port to which the Arduino is connected, you can use Udev rules. Follow these steps:

1. Connect Arduino to the USB port of the computer. This will generate a device file in the /dev folder.

2. Open a terminal.

3. Navigate to /dev folder using this command:

   cd /dev

4. List the device names using this command (see Figure 6-16):

   ls -l | grep -i -E "tty[a-z]"

```
rajesh@ubuntu:/dev$ ls -l | grep -i -E "tty[a-z]"
crwxrw-rwx+ 1 root    dialout 166,   0 Feb 28 23:27 ttyACM0
crw-------  1 root    root      5,   3 Feb 28 23:27 ttyprintk
crw-rw----  1 root    dialout   4,  64 Feb 28 23:27 ttys0
crw-rw----  1 root    dialout   4,  65 Feb 28 23:27 ttys1
crw-rw----  1 root    dialout   4,  74 Feb 28 23:27 ttys10
crw-rw----  1 root    dialout   4,  75 Feb 28 23:27 ttys11
crw-rw----  1 root    dialout   4,  76 Feb 28 23:27 ttys12
crw-rw----  1 root    dialout   4,  77 Feb 28 23:27 ttys13
crw-rw----  1 root    dialout   4,  78 Feb 28 23:27 ttys14
crw-rw----  1 root    dialout   4,  79 Feb 28 23:27 ttys15
crw-rw----  1 root    dialout   4,  80 Feb 28 23:27 ttys16
crw-rw----  1 root    dialout   4,  81 Feb 28 23:27 ttys17
```

Figure 6-16. List of device names

5. Identify the device name of Arduino (for example, ttyACM0). This can be done by disconnecting and reconnecting Arduino to the computer and listing the devices. The device name is the one that disappears while disconnecting Arduino and appears when reconnecting it.

6. Get the device information by typing the following command:

```
udevadm info /dev/ttyACM0 | grep -E 'ID_
VENDOR_ID|ID_MODEL_ID|ID_SERIAL_SHORT'
```

Where ttyACM0 is the name of the device file of Arduino. This command will display the vendor ID, model ID, and serial number, as shown in Figure 6-17.

Figure 6-17. *Device information displayed using the udevadm command*

This device information allows you to identify the Arduino each time it is plugged into the computer.

7. Next, you need to create a rules file containing the device information and assign a label to the Arduino of your choice. To do this, type the following commands:

```
cd /etc/udev/rules.d/
sudo gedit my_rules.rules
```

Now type the following details into the document and save it:

```
SUBSYSTEM=="tty", SUBSYSTEMS=="usb",
ATTRS{idVendor}=="2341",
ATTRS{idProduct}=="0043", ATTRS{serial
}=="75630313636351A052C1", MODE="0777",
SYMLINK+="arduino_uno"
```

Subsequently, assign values to the idVendor, and idProduct variables and the serial number from the device information you retrieved in Step 6. Here, MODE=0777 assigns read/write and execute permissions to the device and SYMLINK+="arduino_uno" signifies the assigning of a custom label to the device (here, arduino_uno is the custom label).

8. Next, type the following command to have the rules take effect:

```
sudo udevadm control --reload-rules && sudo
service udev restart && sudo udevadm trigger
```

9. To verify whether the rules have taken effect, type the following command:

```
1. cd /dev
2. ls -l | grep arduino
```

If everything went well, you will see output similar to that shown in Figure 6-18.

```
rajesh@ubuntu:~$ cd /dev
rajesh@ubuntu:/dev$ ls -l | grep arduino
lrwxrwxrwx  1 root    root            7 Mar  1 08:12 arduino_uno -> ttyACM0
```

Figure 6-18. *Output of the command*

Here, you can see that the device is assigned the name arduino_uno, and read/write/execute permissions are assigned (indicated by rwx on the left). Also, whenever you unplug and then plug the device back in, the name along with the permissions are reassigned properly, as per your rules.

Arduino and ROS Integration Examples

This section illustrates examples of connecting Arduino code with ROS using publishers/subscribers and services.

LED Blink Using ROS Publisher-Subscriber

Here is the code:

```
1.  //LED Blink - ROS
2.
3.  #include <ros.h>
4.  #include <std_msgs/Empty.h>
5.
6.  //Macros
7.  #define pin_number 5
8.
9.  //ROS NodeHandle
10. ros::NodeHandle nh;
11.
12. //Global Variables
13. int count = 0, no_of_times = 5, wait_time = 500;
14.
15. void blink_led()
16. {
17.   while(true)
18.   {
19.     if (count < no_of_times)
20.     {
21.       digitalWrite(pin_number, HIGH);
22.       delay(wait_time);
23.       digitalWrite(pin_number, LOW);
```

```
24.        delay(wait_time);
25.        count++;
26.      }
27.    else
28.    {
29.      break;
30.    }
31.  }
32. }
33.
34. void messageCb( const std_msgs::Empty& toggle_msg)
35. {
36.   count = 0;
37.   blink_led();
38. }
39.
40. ros::Subscriber<std_msgs::Empty> sub("toggle_led",
    &messageCb );
41.
42. void setup()
43. {
44.   //Debug
45.   Serial.begin(57600);
46.
47.   //LED
48.   pinMode(pin_number, OUTPUT);
49.
50.   //ROS NodeHandle
51.   nh.initNode();
52.
53.   //ROS Subscriber
```

```
54.   nh.subscribe(sub);
55. }
56.
57. void loop()
58. {
59.   nh.spinOnce(); //Waiting
60.   delay(1);
61. }
```

This Arduino code creates a ROS node that subscribes to a topic
named toggle led. When the node receives a message, it toggles an LED a
certain number of times with a predetermined delay between each toggle.
The code works as follows:

- Line 3 imports the library named ros.h, which allows
  you to create nodes, define publishers/subscribers/
  services, and so on.

- Line 4 imports another library named std_msgs/
  Empty.h, which defines the std_msgs/Empty
  message type.

- Line 7 of the program creates a macro named pin_
  number to control the LED.

- Line 10 defines a ROS NodeHandle object, which creates
  publishers, subscribers, and so on.

- After that, it creates and initializes the following
  variables:

  - count: Keeps track of how many times the LED has
    been toggled

  - no_of_times: Indicates the number of times the
    toggling should be done

317

- `wait_time`: Denotes the delay between the LED toggle (500 milliseconds)

- The logic for switching the LED is contained in the `blink_led()` method. It does the following:

  - The LED is switched on and off using a `while` loop. It continues until the `count` variable is more than or equal to the predetermined number of times.

  - Using the `delay()` method and the `digitalWrite()` function, the program turns the LED on and off and delays a certain amount of time inside the loop.

  - Each time a toggle happens, the `count` variable is increased.

- The callback method, `messageCb()`, is called whenever a message on the `toggle led` topic is received. It invokes the `blink_led()` method to turn the LED on and off for a predetermined number of times after setting the `count` variable to 0.

- Line 40 creates a ROS subscriber called `sub` that subscribes to the `toggle_led` topic and sets its callback function to `messageCb()`.

- The `setup()` function does the following:

  - Sets up the serial communication.

  - Initializes the ROS node using the `nh. initNode()` method.

  - Initializes the LED pin mode to `OUTPUT`.

  - Subscribes to the `toggle led` topic.

- The program then enters the loop() method, where:

  - The nh.spinOnce() function keeps the node active and receptive to incoming messages on the toggle_led topic.

  - The delay() method stops the loop from running too quickly and using up too much CPU power.

LED Trigger Using the ROS Service

Here is the code:

```
1. //LED - ROS Service
2.
3. #include <ros.h>
4. #include <std_srvs/SetBool.h>
5.
6. //Macros
7. #define pin_number 5
8.
9. //ROS NodeHandle
10. ros::NodeHandle  nh;
11.
12. //Service Type
13. using std_srvs::SetBool;
14.
15. void callback(const SetBool::Request & req,
    SetBool::Response & res){
16.   if(req.data)
17.   {
18.     digitalWrite(pin_number, HIGH); //Turn LED ON
19.     res.success = true;
20.     res.message = "LED turned ON";
```

```
21.   }
22.   else
23.   {
24.     digitalWrite(pin_number, LOW); //Turn LED OFF
25.     res.success = false;
26.     res.message = "LED turned OFF";
27.   }
28. }
29.
30. ros::ServiceServer<SetBool::Request, SetBool::Response>
    server("LED_Service",&callback);
31.
32. void setup()
33. {
34.     //Debug
35.   Serial.begin(57600);
36.
37.   //LED
38.   pinMode(pin_number, OUTPUT);
39.
40.   //ROS NodeHandle
41.   nh.initNode();
42.
43.   //ROS Service Server
44.   nh.advertiseService(server);
45. }
46.
47. void loop()
48. {
49.   nh.spinOnce();
50.   delay(10);
51. }
```

This program creates a ROS service called LED Service that can be utilized to turn on/off an LED. The code works as follows:

- Line 3 imports the library named ros.h, which allows you to create nodes, define publishers/subscribers/ services, and so on, in an Arduino program.

- Line 4 imports another library named std_srvs/ SetBool.h, which defines the std_srvs/SetBool service type.

- Line 7 creates a macro for the pin number of the LED you want to control.

- Afterward, it generates a NodeHandle object, which communicates with ROS.

- The SetBool::Request and SetBool::Response parameters are required for the callback() method. Each time a client requests the LED_Service service, this function is invoked. The callback function does the following:

  - If the data field of the request message is true:

    - The LED is turned on.

    - The success field of the response message is set to true.

    - The message field of the response message is set to LED turned ON.

  - If the data field of the request message is false:

    - The LED is turned off.

    - The success field of the response object is set to false.

- • The message field of the response message is set to LED turned OFF.

- • In Line 30, a ROS service server object named server is created. It contains a service called LED_Service. The callback() callback function is also specified and is invoked upon receiving a request.

- • The setup() method does the following:

    - • Initializes the serial communication.

    - • Changes the LED pin mode to OUTPUT.

    - • Initializes the ROS node using the nh.initNode() method.

    - • Advertises LED_Service using the nh.advertiseService() function.

- • The software then enters the loop() method, where:

    - • The nh.spinOnce() function keeps the node active and waits for incoming service requests.

    - • The delay() method stops the loop from running too quickly and using up too much CPU power.

Buzzer Control Using the ROS Publisher-Subscriber

Here is the code:

```
1. //Buzzer - ROS
2.
3. #include <ros.h>
4. #include <std_msgs/Empty.h>
```

```
5. // Buzzer
6. #define pin_number 8
7. int count = 0, no_of_times = 5, note_duration = 0, delay_
   time = 200, freq_start = 100, freq_end = 300;
8.
9. //ROS NodeHandle
10. ros::NodeHandle nh;
11.
12. void generate_sound()
13. {
14.   while(true)
15.   {
16.     if(count<no_of_times)
17.     {
18.       for(int note=freq_start; note<freq_end; note+=25)
19.       {
20.         tone(pin_number, note, note_duration);
21.         delay(delay_time);
22.         noTone(pin_number);
23.       }
24.       count++;
25.     }
26.     else
27.     {
28.       break;
29.     }
30.   }
31. }
32.
33. void messageCb( const std_msgs::Empty& toggle_msg)
34. {
```

```
35.    count = 0;
36.    generate_sound();
37. }
38.
39. ros::Subscriber<std_msgs::Empty> sub("toggle_buzzer",
    &messageCb );
40.
41. void setup()
42. {
43.    //Debug
44.    Serial.begin(57600);
45.
46.    //Buzzer
47.    pinMode(pin_number, OUTPUT); // declare as output pin
48.
49.    //ROS NodeHandle
50.    nh.initNode();
51.
52.    //ROS Subscriber
53.    nh.subscribe(sub);
54. }
55.
56. void loop()
57. {
58.    nh.spinOnce(); //Waiting
59.    delay(1);
60. }
```

This Arduino code creates a ROS node that subscribes to a topic named toggle buzzer. When the node receives a message, it toggles a buzzer a certain number of times, generating frequencies within a range. The code works as follows:

- Line 3 imports the library named ros.h, which allows you to create nodes, define publishers/subscribers/ services, and so on.

- Line 4 imports another library named std_msgs/ Empty.h, which defines the std_msgs/Empty message type.

- Line 6 of the program creates a macro for the pin number of the buzzer you want to control.

- After that, it creates and initializes the following variables:

  - count: Keeps track of how many times the buzzer has been toggled.

  - no_of_times: Indicates the number of times the toggling needs to be done.

  - note_duration: Indicates how long the note needs to be played.

  - delay_time: Denotes the delay between the buzzer toggle (200 milliseconds).

  - freq_start and freq_end: Defines the start and end of the frequency range that needs to be played.

- Line 10 creates a ROS NodeHandle object, which creates publishers, subscribers, and so on.

- The logic for generating frequencies is contained in the generate_sound() method. It does the following:

  - The buzzer is triggered using a while loop that continues until the count variable equals or exceeds the value of the no of times variable.

- The note variable is incremented from freq_start to freq_end with a step of 25, in the for loop.

- The tone() method creates a tone. It has the following parameters:

  - Frequency of the sound

  - Designated pin number of the buzzer

  - Note duration

- The noTone() method switches off the tone after a delay time.

- The callback method, messageCb(), is called whenever a message on the toggle_buzzer topic is received. It invokes the generate_sound() method to generate frequencies on the buzzer, after setting the count variable to 0.

- Line 39 creates a ROS subscriber called sub that subscribes to the toggle_buzzer topic and sets the callback function to messageCb().

- The setup() function does the following:

  - Sets up the serial communication.

  - Initializes the ROS node using the nh. initNode() method.

  - Initializes the LED pin mode to OUTPUT.

  - Subscribes to the toggle led topic.

- The program then enters the loop() method, where:

- The nh.spinOnce() function keeps the node active and receptive to incoming messages on the toggle_buzzer topic.

- The delay() method stops the loop from running too quickly and using up too much CPU power.

Switch Control Using the ROS Publisher-Subscriber

Here is the code:

```
1. //Switch-ROS
2. #include <ros.h>
3. #include <std_msgs/Bool.h>
4.
5. //ROS NodeHandle
6. ros::NodeHandle nh;
7.
8. //Message
9. std_msgs::Bool switch_msg;
10.
11. //Publisher
12. ros::Publisher pub_switch("switch_status", &switch_msg);
13.
14. //Macros
15. #define switch_pin 3
16.
17. //Global Variables
18. int val = 0; // variable for reading the pin status
19.
20. void setup()
```

```
21. {
22.    //Debug
23.    Serial.begin(57600);
24.
25.    //ROS NodeHandle
26.    nh.initNode();
27.
28.    //ROS Publisher
29.    nh.advertise(pub_switch);
30.
31.    //Switch
32.    pinMode(switch_pin, INPUT);
33. }
34.
35. void loop()
36. {
37.    val = digitalRead(switch_pin);
38.    if (val == HIGH)
39.    {
40.      switch_msg.data = true;
41.    }
42.    else
43.    {
44.      switch_msg.data = false;
45.    }
46.    pub_switch.publish(&switch_msg);
47.    nh.spinOnce();
48.    delay(10);
49. }
```

This Arduino code creates a ROS node that broadcasts the state of the switch attached to pin 3 of the Arduino board The code works as follows:

- Line 2 imports the library named ros.h, which allows you to create nodes, define publishers/subscribers/ services, and so on.

- Line 3 imports another library named std_msgs/ Bool.h, which defines the std_msgs/Bool message type.

- Line 6 creates a ROS NodeHandle object to create publishers, subscribers, and so on.

- Line 9 creates an object of type std_msgs/Bool to publish the state of the switch.

- Line 12 creates a publisher called pub_switch, which publishes under a topic called switch_status. The &switch_msg parameter refers to the message to be published.

- Line 15 of the program creates a macro for the pin number of the switch you want to read.

- After that, it creates and initializes the global integer variable called val, which contains the switch status (i.e., true or false).

- The setup() function of the program does the following:

  - Sets up the serial communication.

  - Initializes the ROS node using the nh. initNode() method.

- Registers the publisher with the ROS master, making it visible to other nodes.

- Initializes the switch pin mode to INPUT.

- The program then enters the loop() method, where:

  - The digitalRead(switch_pin) method reads the value of the pin connected to the Arduino.

  - If the value of the switch is ON, the true value is assigned to the data field of switch_msg. Otherwise, the false value is assigned.

  - The message is published using the pub_switch. publish(&switch_msg) code line.

  - The nh.spinOnce() function keeps the node active.

  - The delay() method publishes messages at regular intervals and prevents the use of too much CPU power.

LCD Control Using the ROS Publisher-Subscriber

Here is the code:

```
1. //LCD - ROS
2.
3. #include <ros.h>
4. #include <Wire.h>
5. #include <std_msgs/String.h>
6. #include <LiquidCrystal_I2C.h>
7.
8. //ROS NodeHandle
9. ros::NodeHandle nh;
```

```
10.
11. //Global Variables
12. LiquidCrystal_I2C lcd(0x27, 16, 2);
13.
14. //DISPLAY IN LCD
15. void display_lcd(String message)
16. {
17.   lcd.clear();
18.   lcd.setCursor(0,0);
19.   lcd.print(message);
20. }
21.
22. //Subscriber Callback Function - LCD
23. void lcd_callback(const std_msgs::String& lcd_msg)
24. {
25.   display_lcd(lcd_msg.data);
26. }
27.
28. //ROS Variables
29. ros::Subscriber<std_msgs::String> lcd_subscriber("lcd",
    lcd_callback);
30.
31. //Constructor
32. void setup()
33. {
34.   //Debug
35.   Serial.begin(57600);
36.   //LCD
37.   lcd.begin();
38.   lcd.backlight();
39.   lcd.clear();
```

```
40.
41.   //ROS NodeHandle
42.   nh.initNode();
43.
44.   //ROS Publishers and Subscribers
45.   nh.subscribe(lcd_subscriber);
46. }
47.
48. //Main
49. void loop()
50. {
51.   nh.spinOnce();//Waiting
52. }
```

This Arduino code implements a ROS node that subscribes to messages and displays them on an LCD screen. The code works as follows:

- Lines 3 to 6 include the LiquidCrystal I2C library, a header file for ROS, Wire (I2C communication), and the LCD screen interface.

- To enable communication between the node and the ROS system, the global variable nh is created in Line 9 as an object of type ros::NodeHandle.

- Line 12 creates an instance named lcd of type LiquidCrystal_I2C, which takes three parameters:

  - 0x27 specifies the I2C address of the LCD and is the default address of many LCD modules.

  - 16 refers to the number of columns in the LCD.

  - 2 specifies the number of rows in the LCD.

- The `display_lcd(String message)` method does the following:

  - Takes one string value as input.

  - Clears the contents of the LCD screen.

  - Sets the cursor to column 0, row 0 of the display.

  - Prints the string message on the LCD screen.

- A subscriber is created on Line 29 named `lcd_subscriber`. It subscribes to a topic named `lcd`. Also, when a message is received, it invokes a callback function named `lcd_callback()`.

- Lines 23 to 26 define the `lcd` callback function, which is in charge of receiving messages on the `lcd` topic and uses the `display_lcd` function to display them on the LCD screen.

- The `setup()` function does the following:

  - Sets up the serial communication.

  - Initializes the ROS node using the `nh.initNode()` method.

  - Initializes the LCD.

  - Subscribes to the `lcd` topic.

- The program then enters the `loop()` method, where:

  - The `nh.spinOnce()` function keeps the node active and receptive to incoming messages on the `lcd` topic.

  - The `delay()` method stops the loop from running too quickly and using up too much CPU power.

Summary

This chapter explained Arduino basics, Arduino programming basics, and interfacing Arduino with ROS. It included Arduino and ROS integration examples. The next chapter describes the preliminary setup process, including building the robot model (URDF), designing the robot parts using 3D modeling software, adding 3D models inside URDF, visualizing the robot, simulating the robot, teleoperation using the keyboard/joystick/ android devices, mapping, and more.

CHAPTER 7

Simulating Bumblebot: A Simple Two-Wheeled Robot

Outline

This chapter discusses how to build a basic two-wheeled robot named "Bumblebot" in simulation. It consists of the following steps:

- Preliminary setup

- Building the robot model (URDF)

- Designing robot parts using 3D modeling software

- Adding 3D models inside URDF

- Robot visualization

- Robot simulation

© Rajesh Subramanian 2023
R. Subramanian, *Build Autonomous Mobile Robot from Scratch using ROS*,
Maker Innovations Series, https://doi.org/10.1007/978-1-4842-9645-5_7

- Tele-operation

  - Using the keyboard

  - Using a joystick

  - Using an Android device

- Mapping

- Autonomous navigation

- Tuning navigation

Preliminary Setup

To simulate Bumblebot on a computer, you need to perform these preliminary setup procedures:

1. Clone the robot workspace from Git into the computer:

    ```
    git clone https://github.com/logicraju/BumbleBot_WS.git
    ```

2. Go to the robot workspace:

    ```
    cd BumbleBot_WS
    ```

3. In Git, create multiple branches (versions) of the software. Here, the default branch is ROS Noetic. If you're using ROS Melodic, change the Git branch to melodic-devel using the following command:

    ```
    git checkout origin/melodic-devel
    ```

4. Install the required ROS packages and dependencies:

    ```
    rosdep install --from-paths src --ignore-
    src -r -y
    ```

5. Compile the workspace:

```
cd ~/Bumblebot_WS
catkin_make
```

6. Source the workspace:

```
source ~/Bumblebot_WS/devel/setup.bash
```

To permanently source the workspace, add the command to the end of the ~/.bashrc file, as shown in Figure 7-1.

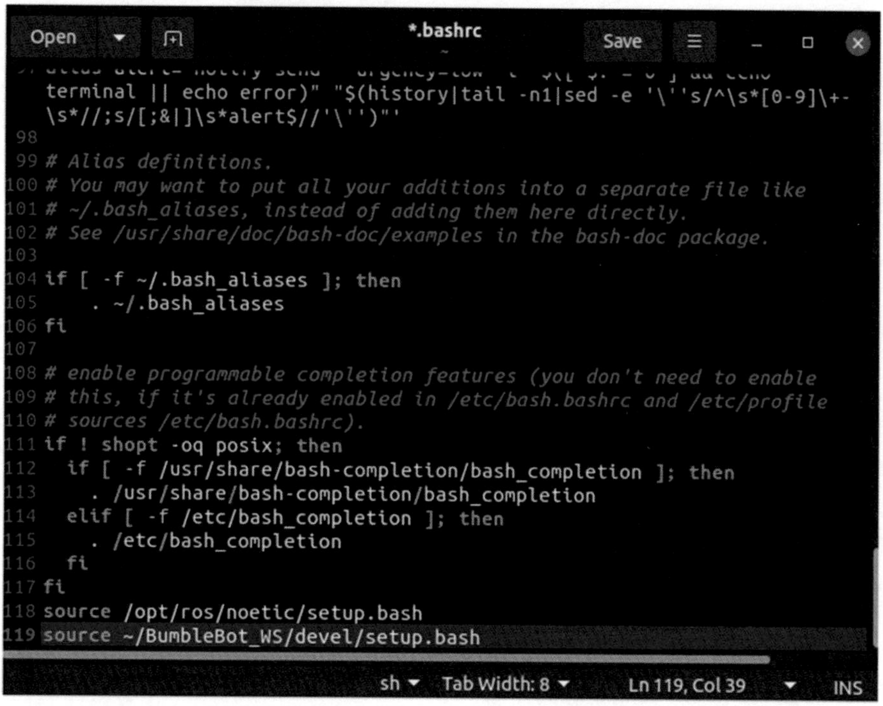

Figure 7-1. *The .bashrc file*

Building the Robot Model (URDF)

In ROS, robot models are defined as links and joints in Unified Robot Description Format (URDF), which is an XML file. Links are the fixed part of the robot and joints are the moveable parts. Joints enable the movement of the links in the robot. There are six types of joints supported in ROS:

- **Revolute**: A hinge joint that rotates along the axis with a limited range of motion. For example, the Elbow joint of a robotic arm.

- **Continuous**: A hinge joint that rotates along the axis without limits. For example, the Wheel joint of a mobile robot.

- **Prismatic**: A sliding joint that slides along the axis and has a limited range of motion. For example, Joints of a cartesian robot.

- **Fixed**: A non-moveable type of joint. For example, a sensor joint attached to the body of a robot (lidar, ultrasonic, camera, etc.).

- **Floating**: Free to rotate along any axis.

- **Planar**: Enables movement perpendicular to the axis.

Note Continuous and fixed type joints are used in Bumblebot.

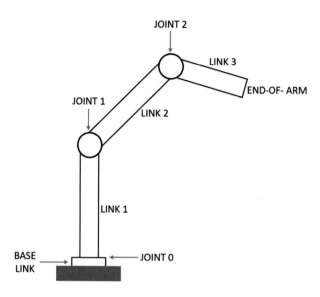

Figure 7-2. *Illustration of a robot arm with three links and two revolute joints*

Figure 7-2 depicts a robot arm that has four links, two revolute joints, and one fixed joint. The four links of the robot arm are the base link, link1, link2, and link3. The base link and link1 are connected via joint0, which is a fixed joint. Link1 and link2 are connected using a revolute joint (joint1). Link2 and link3 are connected using a revolute joint (joint2).

The next section explains the links and joints of Bumblebot.

Links

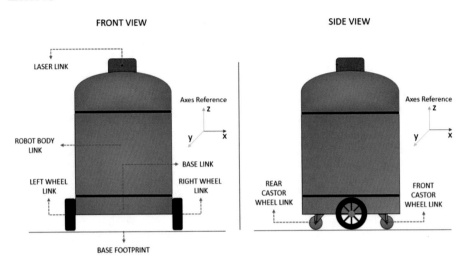

Figure 7-3. Bumblebot links

Bumblebot has eight links, as shown in Figure 7-3. They are as follows:

- **Base footprint**: A virtual representation of the robot on the ground and used as the origin for the coordinate frames of the robot's links and joints. The position and orientation of the robot are calculated using the base footprint link wrt (with respect to) the world. This helps the robot perform complex operations such as navigation, obstacle avoidance, and so on.

- **Base Link**: Unlike the base footprint, the base link is an actual part and is usually located at the geometric center of the robot. It is the root link to which all other links are attached (directly or indirectly).

Note   In Bumblebot, the base link is located at the bottom center instead of the geometric center of the robot.

- **Robot body link**: The main chassis of the robot is attached to the base link.

- **Left wheel link**: The left wheel of the robot is attached to the base link.

- **Right wheel link**: The right wheel of the robot is attached to the base link.

- **Front castor wheel link**: The front castor wheel of the robot is attached to the base link.

- **Rear castor wheel link**: The rear castor wheel of the robot is attached to the base link.

- **Laser link**: Is the Lidar scanner is mounted on top of the robot and is connected to the base link.

Joints

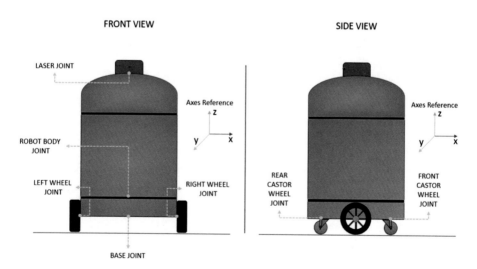

Figure 7-4. *Bumblebot joints*

Bumblebot has seven joints, as shown in Figure 7-4. They are as follows:

- **Base Joint**: A fixed-type joint that connects the base footprint link with the base link.

- **Robot Body Joint**: A fixed-type joint that connects the base link with the robot body link.

- **Left Wheel Joint**: A continuous type joint that connects the left wheel link with the base link.

- **Right Wheel Joint**: A continuous type joint that connects the right wheel link with the base link.

- **Front Castor Wheel Joint**: A continuous type joint that connects the front caster wheel link to the base link.

- **Rear Castor Wheel Joint**: A continuous type joint that connects the rear caster wheel link to the base link.

- **Laser Joint**: A fixed-type joint that connects the laser link with the base link.

The next section explains how to describe the links and joints using the URDF format.

URDF

To describe the kinematics of a robot (i.e., the links and joints), you use a format known as Universal Robot Description Format (URDF). URDF is an XML file with several tags.

Let's look at the important parts of the URDF description of Bumblebot.

```
<?xml version="1.0"?>
```

This is the declaration that specifies the version of XML used in the file.

```
<robot name="robot">
```

This tag defines the name of the robot. All the other tags (such as links, joints, etc.) are specified within this tag.

```
<link name="base_footprint" />
```

This represents the origin of the robot. It does not have any geometry or mass properties and is typically used to define the coordinate system for the robot.

```
<material name="Orange">
  <color rgba="0.5 0.2 0.0 1" />
</material>
```

This defines a color named "Orange" with a specific RGBA (red, blue, green, alpha/transparency) color. RGBA values range from 0 to 1. Other colors are also defined in a similar way including yellow, red, blue, and green. These colors are applied to the links of the robot for visualization purposes. Colors are specified within the <material> tag.

```
<link name="base_link">
  <inertial>
    <origin xyz="0 0 0" rpy="0 0 0" />
    <mass value="70.0" />
    <inertia ixx="0.079945" ixy="0.00012078" ixz="0.00015606"
    iyy="0.068406" iyz="-0.0049053" izz="0.12343" />
  </inertial>
<visual>
  <!--origin xyz="0 0 0" rpy="0 0 -1.57" /-->
  <origin xyz="0 0 -0.015" rpy="0 0 -1.57" />
  <geometry>
    <!--cylinder length="0.007" radius="0.13"/-->
    <mesh filename="package://my_robot_model/urdf/meshes/
    Base_Plate.stl" scale="0.001 0.001 0.001"/>
  </geometry>
```

```
  <material name="Blue" />
</visual>
<collision>
  <!--origin xyz="0 0 0" rpy="0 0 -1.57" /-->
  <origin xyz="0 0 -0.015" rpy="0 0 -1.57" />
  <geometry>
    <!--cylinder length="0.007" radius="0.13"/-->
    <mesh filename="package://my_robot_model/urdf/meshes/
    Base_Plate.stl" scale="0.001 0.001 0.001"/>
  </geometry>
</collision>
</link>
```

This block defines a link called base_link with some properties. To define a link, you need to specify three main properties. They are inertial, visual, and collision properties.

- The <inertial> tag defines the physical properties of the link such as mass, the center of mass, and inertia. These values are required to simulate the robot in Gazebo.

  - The <origin> tag specifies the center of mass of the link. It is represented as coordinates (x, y, z) to denote position and angles (roll, pitch, yaw) to denote orientation.

  - The <mass> tag specifies the mass of the link in kilograms.

  - The <inertia> tag specifies the moment of inertia matrix. That is, the spatial distribution of mass.

- The <visual> tag describes the visual elements of a link, such as shape, color, and geometric center.

- The <origin> tag specifies the geometric center
  of the shape. It is given as coordinates (x, y, z) and
  angles (roll, pitch, yaw).

- The <geometry> tag defines the shape of the link.
  You can either define the shape by using basic
  geometric figures such as cylinder, box, sphere, and
  so on, or by loading a 3D model file.

- The <material> tag adds color to the shape.

- The <collision> tag helps detect collisions of the link
  with other objects during simulation. For this, you need
  to define the boundaries of the link. This is done using
  the following tags:

  - The <origin> is typically the same as the geometric
    center of the shape mentioned in the <visual> tag.

  - The <geometry> tag specifies the boundaries of the
    link and detects collisions. Usually, the shape of the
    link mentioned in the <geometry> tag (inside the
    <visual> tag) is used.

```
<joint name="base_joint" type="fixed">
  <origin xyz="0 0 0.0335" rpy="0 0 0" />
  <parent link="base_footprint" />
  <child link="base_link" />
  <axis xyz="0 0 0" />
</joint>
```

This section defines a joint named base_joint, which is a fixed joint type that connects the base_footprint link to the base_link link. To define a joint, we need to specify four main parameters. They are as follows:

- The <origin> tag defines the origin of the joint. That is, the point where the two links are connected. You need to specify the position and orientation of the joint in the origin tag as coordinates (x, y, z) and angles (roll, pitch, yaw).

- The <parent> tag specifies the name of the parent link. Here, base_footprint is the parent link.

- The <child> tag specifies the name of the child link. Here, base_link is the child link.

- The <axis> tag defines the axis of rotation of the joint. Here, the joint does not move as it is a "fixed" type.

```
<joint name="right_wheel_joint" type="continuous">
  <origin xyz="0.0 -0.12 0.01" rpy="1.571 0 0" />
  <parent link="base_link" />
  <child link="right_wheel_link" />
  <axis xyz="0 0 -1" />
</joint>
```

Let's look at another joint, which is a "continuous" type. This represents the right wheel joint and enables the wheel to rotate along the negative z-axis (specified in the <axis> tag) of the joint without any limits.

```
<transmission name="right_wheel_trans" type="SimpleTransmission">
  <type>transmission_interface/SimpleTransmission</type>
  <actuator name="right_wheel_motor">
    <mechanicalReduction>1</mechanicalReduction>
```

```
  </actuator>
  <joint name="right_wheel_joint">
    <hardwareInterface>VelocityJointInterface</hardwareInterface>
  </joint>
</transmission>
```

The `<transmission>` tag contains information about the type of transmission, such as whether it is a gearbox, a differential, or a simple transmission. It also includes information about the actuator that drives the transmission, such as the type of actuator, its joint, and its transmission ratio. In this example, you can see that the `SimpleTransmission` type is used and the mechanical reduction is 1, which means that the output of the transmission is the same as the input from the actuator.

The `<hardwareInterface>` tag specifies the hardware interface of a robot's joint. The hardware interface defines how the joint is controlled, such as through position, velocity, effort, or a combination of these. In this example, you can see that the joint is controlled using velocity.

```
<gazebo reference="base_link">
  <material>Gazebo/Blue</material>
</gazebo>
```

To specify the color of the links of the robot during simulation, you can use the `<material>` tag within the `<gazebo>` tag. This example assigns blue to the base link of the robot.

```
<gazebo>
  <plugin name="differential_drive_controller"
  filename="libgazebo_ros_diff_drive.so">
    <!-- Plugin update rate in Hz -->
    <updateRate>10</updateRate>
    <!-- Name of left joint, defaults to `left_joint` -->
    <leftJoint>left_wheel_joint</leftJoint>
```

```
<!-- Name of right joint, defaults to `right_joint` -->
<rightJoint>right_wheel_joint</rightJoint>
<!-- The distance from the center of one wheel to the
other, in meters, defaults to 0.34 m -->
<wheelSeparation>0.23</wheelSeparation>
<!-- Diameter of the wheels, in meters, defaults to
0.15 m -->
<wheelDiameter>0.085</wheelDiameter>
<!-- Wheel acceleration, in rad/s^2, defaults to 0.0
rad/s^2 -->
<wheelAcceleration>0.75</wheelAcceleration>
<!-- Maximum torque which the wheels can produce, in
Nm, defaults to 5 Nm -->
<wheelTorque>500</wheelTorque>
<!-- Topic to receive geometry_msgs/Twist message
commands, defaults to `cmd_vel` -->
<commandTopic>cmd_vel</commandTopic>
<!-- Topic to publish nav_msgs/Odometry messages,
defaults to `odom` -->
<odometryTopic>odom</odometryTopic>
<!-- Odometry frame, defaults to `odom` -->
<odometryFrame>odom</odometryFrame>
<!-- Robot frame to calculate odometry from, defaults
to `base_footprint` -->
<robotBaseFrame>base_footprint</robotBaseFrame>
<!-- Odometry source, 0 for ENCODER, 1 for WORLD,
defaults to WORLD -->
<odometrySource>1</odometrySource>
<!-- Set to true to publish transforms for the wheel
links, defaults to false -->
<publishWheelTF>true</publishWheelTF>
```

```
    <!-- Set to true to publish transforms for the
    odometry, defaults to true -->
    <publishOdom>true</publishOdom>
    <!-- Set to true to publish sensor_msgs/JointState on /
    joint_states for the wheel joints, defaults to false -->
    <publishWheelJointState>true</publishWheelJointState>
    <!-- Set to true to swap right and left wheels,
    defaults to true -->
    <legacyMode>false</legacyMode>
  </plugin>
</gazebo>
```

This example specifies a Gazebo plugin called `differential_drive_controller`, which controls a differential drive robot in simulation.

The parameters of `differential_drive_controller` are as follows:

- **updateRate**: The frequency at which the plugin should be updated in Hz.

- **leftJoint**: The name of the joint that controls the left wheel of the robot.

- **rightJoint**: The name of the joint that controls the right wheel of the robot.

- **wheelSeparation**: The distance between the two wheels of the robot, in meters.

- **wheelDiameter**: The diameter of the robot wheels, in meters.

- **wheelAcceleration**: The maximum acceleration that the wheels can achieve, in radians per second squared.

- **wheelTorque**: The maximum torque that can be applied to the wheels, in Newton meters.

- **commandTopic**: The topic on which the plugin listens for velocity commands. The default topic name is `cmd_vel`, but it can be changed by setting this parameter.

- **odometryTopic**: The topic on which the plugin publishes odometry information. The default topic name is odom, but it can be changed by setting this parameter.

- **odometryFrame**: The frame where the odometry information is expressed. The default frame name is odom, but it can be changed by setting this parameter.

- **robotBaseFrame**: The frame wrt the odometry information is calculated. The default frame name is `base_footprint`, but it can be changed by setting this parameter.

- **odometrySource**: The source of the odometry information. It can be either `ENCODER` or `WORLD`. The default source is `WORLD`.

- **publishWheelTF**: Whether to publish the transform of the wheel links. The default value is `false`.

- **publishOdom**: Whether to publish the odometry information. The default value is `true`.

- **publishWheelJointState**: Whether to publish the state of the wheel joints. The default value is `false`.

- **legacyMode**: Whether to swap the right and left wheels. The default value is `true`.

```
<gazebo reference="laser_link">
  <sensor type="gpu_ray" name="laser">
  <!--sensor type="ray" name="laser"-->
```

```
<pose>0 0 0 0 0 0</pose>
<visualize>false</visualize>
<update_rate>40</update_rate>
<ray>
    <scan>
        <horizontal>
            <samples>720</samples>
            <resolution>1</resolution>
            <min_angle>0</min_angle>
            <max_angle>6.28</max_angle>
        </horizontal>
    </scan>
    <range>
        <min>0.1</min>
        <max>6</max>
        <resolution>0.1</resolution>
    </range>
</ray>
<plugin name="gpu_laser" filename="libgazebo_ros_gpu_
laser.so">
    <topicName>/scan</topicName>
    <frameName>laser_link</frameName>
</plugin>
    </sensor>
</gazebo>
```

This is a Gazebo plugin for a Lidar scanner, which generates scan data of its surroundings. The plugin is specified with the following parameters:

- The `<gazebo reference="laser_link">` tag specifies that the plugin is applied to the link called `laser_link` and mentioned in the URDF.

- The `<sensor type="gpu_ray" name="laser">` tag
  denotes that the sensor is of type `gpu_ray` and the
  name of the sensor is `laser`. You can also use a non-
  GPU version of the laser plugin by changing two lines
  of URDF as shown here:

  - Replace `<sensor type="gpu_ray" name="laser">`
    with `<sensor type="ray" name="laser">`

  - Replace `<plugin name="gpu_laser"`
    `filename="libgazebo_ros_gpu_laser.so">` with
    `<plugin name="laser" filename="libgazebo_`
    `ros_laser.so">`

- The `<pose>` tag specifies the pose (position and
  orientation) of the sensor.

- The `<visualize>` tag is set to `false`, which means that
  the sensor's visualization won't be displayed in the
  Gazebo simulation.

- The `<update_rate>` tag specifies the frequency at
  which the sensor should update its data. In this case,
  it's set to 40Hz.

- The `<ray>` tag specifies the properties of the ray that the
  sensor emits.

  - The `<horizontal>` tag indicates that the laser will
    be emitted in a horizontal plane.

  - The `<samples>720</samples>` tag means that Lidar
    will emit 720 rays or beams.

  - The `<resolution>` tag indicates that the beams will
    be emitted with an angular difference of 1 degree.

- The `<min_angle>` and `<max_angle>` tags specify the minimum and maximum angles in which the laser beams are emitted. Here, the lidar emits beams from 0 to 6.28 radians (i.e., 0 to 360 degrees).

- The `<range>` tag specifies the properties of the range of the sensor.

  - The `<min>` tag specifies the minimum distance that the Lidar can detect obstacles. Here, this means that the sensor will not detect objects that are closer than 0.1 meters.

  - The `<max>` tag specifies the maximum distance that the Lidar can detect obstacles. Here, this means that the sensor will not detect objects that are farther than six meters.

  - The `<resolution>` tag specifies the resolution of the Lidar scanner. Here, it means that the sensor can detect objects with a minimum separation of 0.1 meters. A higher resolution means that the sensor can detect smaller objects or object features, but this comes at the cost of increased computational requirements.

- The `<plugin name="gpu_laser" filename="libgazebo_ros_gpu_laser.so">` tag specifies the name of the plugin and the library filename.

- The `<topicName>` tag specifies the name of the ROS topic on which the scan data will be published. Here, the topic name is `/scan`.

- The `<frameName>` tag specifies the name of the frame where the scan data will be published. Here, the frame name is `laser_link`.

Designing Robot Parts Using 3D Modeling Software

3D modeling is the process of creating a three-dimensional representation of an object. You can use 3D modeling to design robot links, joints, and so on, and use the model in simulation. This allows for better visualization and simulation of the robot when compared to using basic shapes available in URDF such as cylinders, boxes, spheres, and so on. To 3D model an object, you generally begin with basic shapes and modify them to suit your needs. For example, to design a wheel, you can start with a cylinder shape, remove unwanted areas, smooth the edges, place screw holes, and so on. Figure 7-5 shows an example, where different parts of a robot are modeled separately and placed together using the 3D modeling software called Tinkercad.

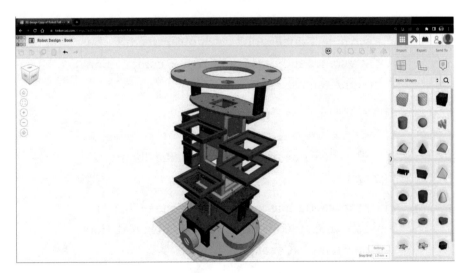

Figure 7-5. *Different links of Bumblebot modeled using Tinkercad*

There are a variety of 3D modeling software programs available on the market. Some of the popular 3D modeling tools include Tinkercad, Fusion 360, SolidWorks, FreeCAD, Blender, and Catia. Tinkercad is an easy and beginner-friendly 3D modeling software. It is available online as a free-to-use web application for 3D modeling. This project used Tinkercad. All the robot parts can be downloaded for 3D printing from this the Git link:

```
https://github.com/logicraju/ROS_Book_WS/blob/main/src/
chapter6/Bumblebot%203d%20Models.zip
```

Using the 3D Models in URDF

After you model the robot, you can use it for simulation and visualization by one of the two methods. In the first method, you can directly export the complete robot model into a format known as Unified Robot Description Format (URDF). 3D modeling tools such as Fusion 360 have a built-in tool to convert the robot model into the URDF format. In the second method, you need to create a basic URDF of the robot, then export the 3D models of the robot parts separately and include them in the URDF file manually.

In the URDF of Bumblebot, a URDF with basic shapes (i.e., cylinders, boxes, etc.) was created. Next, the robot parts were modeled separately using Tinkercad and the basic shapes were replaced with the 3D models in the URDF. Here is the URDF section that uses a 3D model to represent a link.

```
<link name="base_link">
  <inertial>
    <origin xyz="0 0 0" rpy="0 0 0" />
    <mass value="70.0" />
    <inertia ixx="0.079945" ixy="0.00012078" ixz="0.00015606"
    iyy="0.068406" iyz="-0.0049053" izz="0.12343" />
  </inertial>
  <visual>
```

```
    <origin xyz="0 0 -0.015" rpy="0 0 -1.57" />
    <geometry>
      <mesh filename="package://my_robot_model/urdf/meshes/
      Base_Plate.stl" scale="0.001 0.001 0.001"/>
    </geometry>
    <material name="Blue" />
  </visual>
  <collision>
    <origin xyz="0 0 -0.015" rpy="0 0 -1.57" />
    <geometry>
      <mesh filename="package://my_robot_model/urdf/meshes/
      Base_Plate.stl" scale="0.001 0.001 0.001"/>
    </geometry>
  </collision>
  </link>
```

In this code, you can see these lines (which are highlighted):

```
<mesh filename="package://my_robot_model/urdf/meshes/Base_
Plate.stl" scale="0.001 0.001 0.001"/>
```

They specify the location of the 3D model file called Base_Plate.
stl, which represents the base link of the robot. The dimensions of the
3D model are represented in millimeters, and the unit of length in URDF
is specified in meters. Therefore, you need to divide the values in the 3D
model by 1000. The scaling values are given as (0.001, 0.001, 0.001) for (x,
y, and z) values, respectively.

The 3D file is specified in the <visual> and <collision> tags to
visually represent the link and for collision detection purposes during
simulation.

Robot Visualization

The robot kinematic description or URDF is viewed in the visualization tool provided by ROS, called RViz. RViz stands for "ROS Visualization" and helps you view the robot so you can see what it is seeing and doing. For example, this includes the Lidar scan, path planned, goal location, odometry, and so on.

To visualize Bumblebot in RViz, open a terminal and enter the following command:

```
roslaunch bringup visualize.launch
```

Here, the `roslaunch` command launches the `bringup` package, which is the name of the ROS package where the launch file is located, and `visualize.launch` is the name of the launch file. A launch file in ROS starts multiple nodes or programs. Here are the contents of the visualize.launch file:

```
<?xml version="1.0"?>
<launch>
  <!-- Load the robot model into the parameter server -->
  <arg name="urdf_file" default="$(find xacro)/xacro --inorder
  '$(find my_robot_model)/urdf/my_robo_simulation.urdf'" />
  <param name="robot_description" command="$(arg urdf_file)" />

  <!-- Joint State Publisher - Publishes Joint Positions -->
  <node name="joint_state_publisher" pkg="joint_state_
  publisher" type="joint_state_publisher"/>

  <!-- Robot State Publisher - Uses URDF and Joint States to
  compute Forward Kinematics as Transforms -->
  <node name="robot_state_publisher" pkg="robot_state_
  publisher" type="robot_state_publisher"/>
```

```
<!-- RViz - For Visualization -->
<node name="rviz" pkg="rviz" type="rviz" args="-d $(find my_
robot_model)/rviz/my_robo.rviz"/>
```

```
</launch>
```

This code shows the contents of the `visualize.launch` file. The launch file is written in XML. The important parts of the launch file are explained next:

```
<?xml version="1.0"?>
```

This line specifies the XML version. Here, this is 1.0.

```
<launch>
  ...
</launch>
```

This block represents the start and end of the launch file contents.

```
<arg name="urdf_file" default="$(find xacro)/xacro --inorder
'$(find my_robot_model)/urdf/my_robo_simulation.urdf'" />
  <param name="robot_description" command="$(arg urdf_file)" />
```

Here, the robot description mentioned in the URDF is loaded into the parameter server. The parameter server stores the global data, which is accessible to all ROS nodes.

```
<node name="joint_state_publisher" pkg="joint_state_publisher"
type="joint_state_publisher"/>
```

Here, the joint state publisher node runs. The joint state publisher publishes the positions of all the moveable joints in the robot. In Bumblebot, the joint state publisher publishes the positions of left_wheel_joint, right_wheel_joint, front_castor_wheel_joint, and rear_castor_wheel_joint.

```
<node name="robot_state_publisher" pkg="robot_state_
publisher" type="robot_state_publisher"/>
```

Here, the robot state publisher node runs. The robot state publisher takes two inputs: the robot description from the URDF and the updated joint state values. Then, it computes forward kinematics and publishes the transforms. Forward kinematics calculates the new position and orientation based on the updated joint values of the robot.

```
<node name="rviz" pkg="rviz" type="rviz" args="-d $(find my_
robot_model)/rviz/my_robo.rviz"/>
```

Finally, RViz runs. It loads a previously saved RViz file called my_robo. rviz, which displays the robot model.

If everything goes well, you'll see the output shown in Figure 7-6.

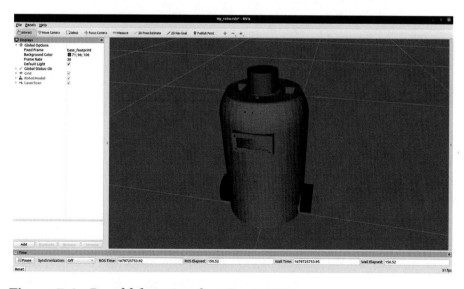

Figure 7-6. *Bumblebot visualization in RViz*

On the left side of the RViz window is the Displays window. You can view all the links of the robot by expanding the RobotModel --> Links menu option (see Figure 7-7).

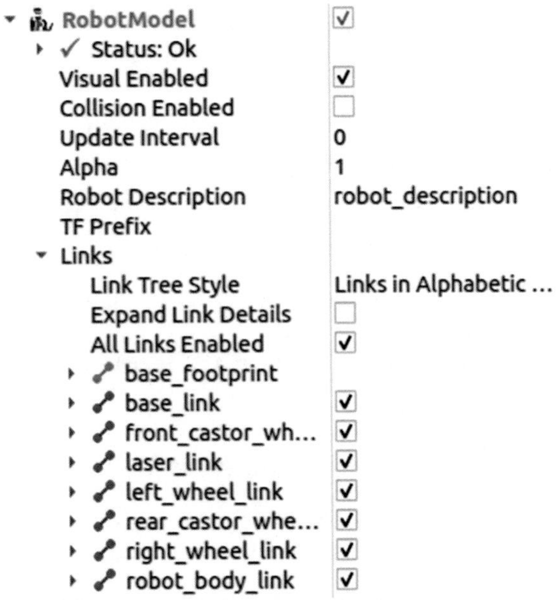

Figure 7-7. Robot's links

Robot Simulation

In this section, you learn how to simulate the robot in the Gazebo simulator. Gazebo approximately represents the behavior of the robot in a virtual world. This allows you to understand how the robot works before you build it. You can virtually add sensors, specify the driving mechanism, and so on, using plugins in Gazebo. Sensor plugins include Lidar, camera, and so on, and drive systems include differential drive, holonomic, skid steer, Ackermann, and so on. Gazebo plugins are blocks of code that can be added to the URDF of the robot. Bumblebot uses the Lidar and Differential Drive plugins, as mentioned in the "URDF" section of this chapter.

To simulate Bumblebot in Gazebo, open a new terminal and enter the following command:

```
roslaunch bringup visualize_gazebo.launch
```

Here, the `roslaunch` command launches the `bringup` package, which is the name of the ROS package where the launch file is located, and `visualize_gazebo.launch` is the name of the launch file. A launch file in ROS starts multiple nodes or programs. Here are the contents of the `visualize_gazebo.launch` file.

```xml
<?xml version="1.0"?>
<launch>

  <!-- Load the robot model into the parameter server -->
  <arg name="urdf_file" default="$(find xacro)/xacro --inorder
'$(find my_robot_model)/urdf/my_robo_simulation.urdf'" />
  <param name="robot_description" command="$(arg urdf_file)" />

  <!-- Launch the gazebo world -->
  <include file="$(find gazebo_ros)/launch/empty_world.launch">
    <arg name="world_name" value="$(find my_robot_model)/
gazebo_worlds/plaza_world.world"/>
  </include>

  <!-- Load the robot model in the parameter server into the
  gazebo world -->
  <!--node name="urdf_spawner" pkg="gazebo_ros" type="spawn_
model" respawn="false" output="screen" args="-urdf -model
my_robo -param robot_description"/-->
  <node pkg="gazebo_ros" type="spawn_model" name="spawn_urdf"
args="-urdf -model my_robo -param robot_description" />

  <!-- Joint State Publisher - Publishes Joint Positions -->
  <node name="joint_state_publisher" pkg="joint_state_
publisher" type="joint_state_publisher"/>

  <!-- Robot State Publisher  - Uses URDF and Joint States to
  compute Transforms -->
```

```
<node name="robot_state_publisher" pkg="robot_state_
publisher" type="robot_state_publisher"/>

<!-- RViz - For Visualization -->
<node name="rviz" pkg="rviz" type="rviz" args="-d $(find my_
robot_model)/rviz/my_robo.rviz"/>

</launch>
```

This code shows the contents of the visualize_gazebo.launch file.
The launch file is written in XML. The important parts of the launch file are
explained next:

```
<?xml version="1.0"?>
```

This line specifies the XML version. Here, this is 1.0.

```
<launch>
   ...
</launch>
```

This block represents the start and end of the launch file contents.

```
<arg name="urdf_file" default="$(find xacro)/xacro --inorder
'$(find my_robot_model)/urdf/my_robo_simulation.urdf'" />
   <param name="robot_description" command="$(arg urdf_file)" />
```

Here, the robot description mentioned in the URDF is loaded into the
parameter server. The parameter server stores the global data, which is
accessible to all ROS nodes.

```
<include file="$(find gazebo_ros)/launch/empty_world.launch">
   <arg name="world_name" value="$(find my_robot_model)/
   gazebo_worlds/plaza_world.world"/>
</include>
```

This line starts the Gazebo simulator and loads an existing Gazebo world with a maze and wooden walls.

```
<node pkg="gazebo_ros" type="spawn_model" name="spawn_urdf"
args="-urdf -model my_robo -param robot_description" />
```

The robot model in the parameter server spawns into the Gazebo world.

```
<node name="joint_state_publisher" pkg="joint_state_publisher"
type="joint_state_publisher"/>
```

Here, the joint state publisher node runs. The joint state publisher publishes the positions of all the moveable joints in the robot. In Bumblebot, the joint state publisher publishes the positions of left_wheel_joint, right_wheel_joint, front_castor_wheel_joint, and rear_castor_wheel_joint.

```
<node name="robot_state_publisher" pkg="robot_state_
publisher" type="robot_state_publisher"/>
```

Here, the robot state publisher node runs. The robot state publisher takes two inputs: the robot description from the URDF and the updated joint state values. Then, it computes forward kinematics and publishes the transforms. Forward kinematics calculates the new position and orientation based on the updated joint values of the robot.

```
<node name="rviz" pkg="rviz" type="rviz" args="-d $(find my_
robot_model)/rviz/my_robo.rviz"/>
```

Finally, RViz runs and loads a previously saved RViz file named my_robo.rviz, which displays the robot model.

If everything goes well, you'll see the output shown in Figures 7-8 and 7-9.

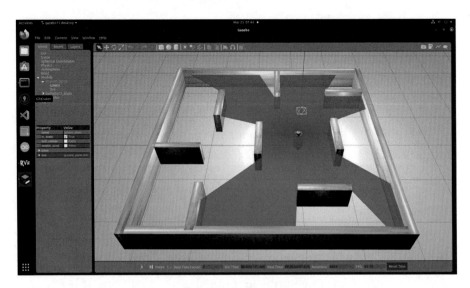

Figure 7-8. *Gazebo simulation of Bumblebot. Blue lines depict laser beams from Lidar*

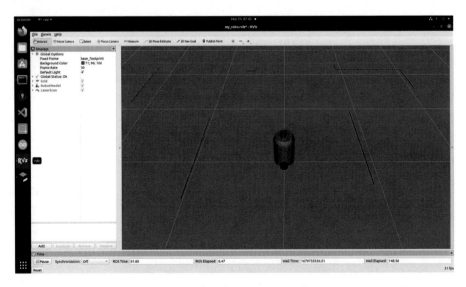

Figure 7-9. *RViz visualization of Bumblebot. Red lines depict the laser scan data*

Teleoperation

Teleoperation involves controlling the movements of the robot using an input device such as a keyboard, joystick, mobile phone, and so on. This includes reading the input keys, computing the linear and rotational velocities depending on the key pressed, publishing the velocities (for the robot as a whole), subscribing the velocities, splitting the velocities into left/right wheel velocities, computing the motor speeds, and generating corresponding voltage to turn the motors.

Now, you see how to teleoperate the Bumblebot using a keyboard, a joystick, and an Android phone.

Teleoperation Using the Keyboard

To teleoperate Bumblebot using a keyboard, open a terminal and enter the following command:

```
roslaunch bringup teleop.launch device:=keyboard
mode:=simulation
```

Here, the `roslaunch` command launches the `bringup` package, which is the name of the ROS package where the launch file is located, and `teleop.launch` is the name of the launch file. A launch file in ROS starts multiple nodes or programs. The `device:=keyboard` and `mode:=simulation` values are command-line arguments. These arguments allow you to run the required launch files and nodes when executing the `roslaunch` command. Here are the contents of the `teleop.launch` file.

```
<?xml version="1.0"?>
<launch>

  <!-- Command Line Argument to Select Mode -->
  <arg name="mode" default="simulation" doc="available modes:
[hardware, simulation]"/>
```

```xml
<!-- Command Line Argument to Select Teleop Device -->
<arg name="device" default="android" doc="available devices:
[keyboard, joystick, android]"/>

<!-- Load the robot model into the parameter server -->
<arg name="urdf_file" default="$(find xacro)/xacro --inorder
'$(find my_robot_model)/urdf/my_robo_$(arg mode).urdf'"/>
<param name="robot_description" command="$(arg urdf_file)" />

<!-- Launch Hardware/Simulation Specific Modules -->
<include file="$(find bringup)/launch/teleop/$(arg mode).
launch.xml"/>

<!-- Joint State Publisher - Publishes Joint Positions -->
<node name="joint_state_publisher" pkg="joint_state_
publisher" type="joint_state_publisher"/>

<!-- Robot State Publisher  - Uses URDF and Joint States to
compute Transforms -->
<node name="robot_state_publisher" pkg="robot_state_
publisher" type="robot_state_publisher"/>

<!-- RVIZ  - Visualization -->
<node name="rviz" pkg="rviz" type="rviz" args="-d $(find my_
robot_model)/rviz/my_robo_teleop.rviz"/>

<!-- Select Teleop Device -->
<include file="$(find bringup)/launch/teleop/$(arg device).
launch.xml">
   <arg name="mode" value="$(arg mode)" />
</include>

</launch>
```

This code shows the contents of the teleop.launch file. The launch file is written in XML. The important parts of the launch file are explained next:

```
<?xml version="1.0"?>
```

This line specifies the XML version. Here, it is 1.0.

```
<launch>
    ...
</launch>
```

This block represents the start and end of the launch file contents.

```
<arg name="mode" default="simulation" doc="available modes:
[hardware, simulation]"/>
```

This shows the command-line argument called mode, which indicates the mode of operation—simulation or hardware. This argument selects the appropriate launch file and URDF file.

```
<arg name="device" default="android" doc="available devices:
[keyboard, joystick, android]"/>
```

This shows the command-line argument called device, which indicates the device used to control the robot—keyboard, joystick, or android. This argument selects the appropriate launch file for teleop.

```
<arg name="urdf_file" default="$(find xacro)/xacro --inorder
'$(find my_robot_model)/urdf/my_robo_$(arg mode).urdf'"/>
<param name="robot_description" command="$(arg urdf_file)" />
```

Here, the robot description mentioned in the URDF is loaded into the parameter server. The parameter server stores the global data, which is accessible to all ROS nodes.

```
<include file="$(find bringup)/launch/teleop/$(arg mode).
launch.xml"/>
```

This runs another launch file, which is selected based on the mode command-line argument.

```
<node name="joint_state_publisher" pkg="joint_state_publisher"
type="joint_state_publisher"/>
```

Here, the joint state publisher node runs. The joint state publisher publishes the positions of all the moveable joints in the robot. In Bumblebot, the joint state publisher publishes the positions of left_wheel_joint, right_wheel_joint, front_castor_wheel_joint, and rear_castor_wheel_joint.

```
<node name="robot_state_publisher" pkg="robot_state_
publisher" type="robot_state_publisher"/>
```

Here, the robot state publisher node runs. The robot state publisher takes two inputs: the robot description from the URDF and the updated joint state values. Then, it computes forward kinematics and publishes the transforms. Forward kinematics calculates the new position and orientation based on the updated joint values of the robot.

```
<node name="rviz" pkg="rviz" type="rviz" args="-d $(find my_
robot_model)/rviz/my_robo_teleop.rviz"/>
```

Here, RViz runs and loads a previously saved RViz file called my_robo_teleop.rviz, which displays the robot model, the Lidar scan, and so on.

```
<include file="$(find bringup)/launch/teleop/$(arg device).
launch.xml">
  <arg name="mode" value="$(arg mode)" />
</include>
```

This runs another launch file, which reads data from the input device and publishes velocity values for the robot. The launch file is selected based on the device command-line argument.

If everything goes well, you will get the output shown in Figure 7-10 and the user will be able to issue commands via the keyboard keys. The keyboard commands must be issued while the cursor is in the terminal where the launch file runs.

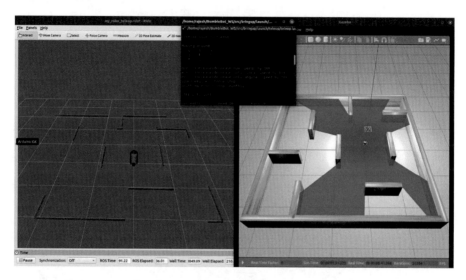

Figure 7-10. *RViz (left), Teleop terminal (middle), and Gazebo (right)*

Teleoperation Using a Joystick

To teleoperate Bumblebot using a joystick, open a terminal and enter the following command:

```
roslaunch bringup teleop.launch mode:=simulation
device:=joystick
```

Here, the `roslaunch` command launches the `bringup` package, is the name of the ROS package where the launch file is located, and `teleop. launch` is the name of the launch file. A launch file in ROS starts multiple nodes or programs. The `device:=joystick` and `mode:=simulation` values are command-line arguments. These arguments allow you to run the required launch files and nodes when executing the `roslaunch` command. The launch file is the same as that used for keyboard teleoperation.

If everything goes well, you'll see the output shown in Figure 7-11 and the user will be able to issue commands via the joystick keys. Unlike the keyboard, the joystick commands need not be issued when the cursor is in the terminal where the launch file runs.

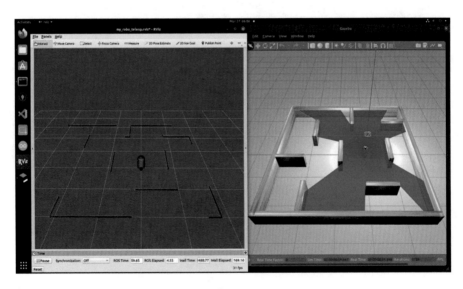

Figure 7-11. *Teleop via a joystick*

Teleoperation Using an Android Device

To teleoperate Bumblebot using an Android device, do the following:

1. Download and install the ROS-Mobile app from the Play Store into your Android device at https://play.google.com/store/apps/details?id=com.schneewittchen.rosandroid.

2. Ensure that the mobile data is turned off.

3. Connect the Android device to the same WiFi network as the robot.

4. Open a terminal (Ctrl+Alt+T).

5. Type hostname -i to display the IP address of the robot.

6. Open the .bashrc file in the home/<user> folder.

7. Add export ROS_MASTER_URI=http://<ip_address>:11311 to the end of the file.

Note   Remove the quotation mark and substitute <ip_address> for the IP address of your robot.

---

8. This command tells all the nodes that the ROS master is running in the specified device.

Now open a terminal (in your computer) and enter the following command:

```
roslaunch bringup teleop.launch mode:=simulation
device:= android
```

Here, the `roslaunch` command launches the `bringup` package, which is the name of the ROS package where the launch file is located, and `teleop.launch` is the name of the launch file. A launch file in ROS starts multiple nodes or programs. The `device:=android` and `mode:=simulation` values are command-line arguments. These arguments allow you to run the required launch files and nodes when executing the `roslaunch` command. The launch file is the same as that used for keyboard and joystick teleoperation.

After launching the simulation, you need to configure the `ROS-Mobile` app on your Android device. Configuration involves the following steps:

1. Go to the Master tab.

2. Type the IP address of the robot (the IP address of the laptop if in simulation) into the Master URI field.

3. Type the port number as 11311 into the Master Port field.

4. Go to the Details tab and select Add Widget.

5. Select Joystick from the drop-down menu.

6. Set the topic name to `cmd_vel`.

7. Go to the Viz menu and control the robot using the virtual joystick.

The configuration steps are shown in Figure 7-12.

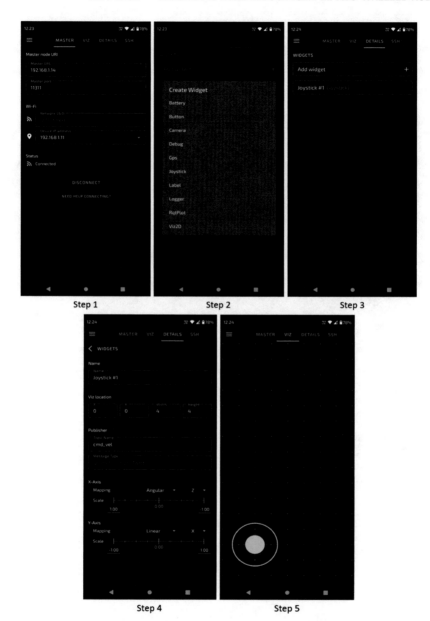

Figure 7-12. *Configuring the ROS-Mobile app for Teleop*

If everything goes well, you'll see the output shown in Figure 7-13 and the user will be able to issue commands via the Android app. Unlike the keyboard, the Android commands need not be issued while the cursor is in the terminal where the launch file runs.

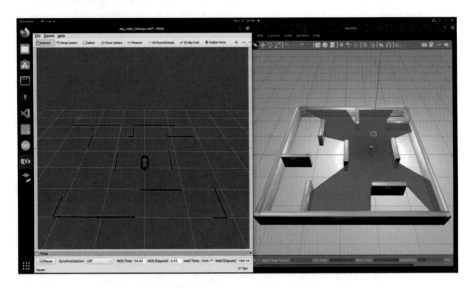

Figure 7-13. *Teleop via Android app*

Mapping

Mapping involves creating a representation of the robot's working environment. A map enables the robot to perform various actions such as localization (estimating the pose of the robot on the map), planning the path toward a goal, performing autonomous navigation, and so on. There are a variety of mapping methods in robotics, including:

- **Occupancy grid mapping**: Involves representing the environment as a grid with several cells. Each cell denotes a small space in the environment. The cells are then marked as occupied or free by using sensor information, as the robot explores the space around it.

- **SLAM**: Stands for Simultaneous Localization and Mapping. As the name indicates, it is a technique to create a map and at the same time estimate the robot's position within the map. In the SLAM algorithm, the robot continuously estimates its position and orientation as it traverses and updates the map, using its sensors. The sensors used for mapping typically involve Lidar, cameras, and odometry sensors (such as wheel encoders, IMU, GPS, etc.).

- **3D mapping**: In this method, a 3D map is generated in contrast to a top-down view obtained from a 2D mapping algorithm (such as occupancy grid mapping and SLAM). The 3D map will include the height and shape of objects in the robot's environment. Sensors used for 3D mapping involve 3D Lidar, stereo cameras, depth cameras, and so on, along with odometry sensors like IMU, wheel encoders, GPS, and so on.

- **Visual SLAM**: This is a variant of the SLAM mapping method, which uses the images obtained from a camera to generate the map. Typical cameras used for visual SLAM are monocular cameras, stereo cameras, and depth cameras. Odometry sensors can also be used along with the cameras for better accuracy.

- **Multi-sensor fusion**: With this method, the data from different sensor sources are combined to create a map for better accuracy. The sensors used can include Lidar, cameras, sonar, and so on.

375

Bumblebot uses a SLAM package provided by ROS called *Gmapping*, using Lidar and wheel encoders to create a map of the environment. The Gmapping algorithm works as follows:

1. The robot's current location is obtained from the odom frame, which is provided by the odometry node.

2. A particle filter tracks the robot's position and orientation (localize) within the map. In the particle filter algorithm, the robot's position is represented by a set of particles, each with a probability weight.

3. Obstacles around the robot are detected by the Lidar.

4. The position of obstacles is calculated with respect to the robot's current location.

5. These obstacle positions are marked on the map.

6. As the robot moves, the algorithm updates the map and the robot's estimated position.

7. After the map has been generated, it can be saved into a file using a node called map_saver, which is provided by the map_server package.

Several parameters in Gmapping can be tuned to optimize map creation. All the parameters can be configured in the gmapping.yaml file in the Gmapping package. For more information on the Gmapping package, see http://wiki.ros.org/gmapping.

To perform mapping on Bumblebot, open a terminal (in your computer) and enter one of the following commands based on the input device of your choice:

```
roslaunch bringup mapping.launch mode:=simulation
device:=keyboard
roslaunch bringup mapping.launch mode:=simulation
device:=joystick
roslaunch bringup mapping.launch mode:=simulation
device:=android
```

Here, the roslaunch command launches the bringup package, which is the name of the ROS package where the launch file is located, and mapping.launch is the name of the launch file. A launch file in ROS starts multiple nodes or programs. The device:=android and mode:=simulation values are command-line arguments that are passed when executing the command. These arguments allow you to run the required launch files and nodes when executing the roslaunch command. Here are the contents of the mapping.launch file.

```
<?xml version="1.0"?>
<launch>

  <!-- Command Line Argument to Select Mode -->
  <arg name="mode" default="simulation" doc="available modes:
[hardware, simulation]"/>

  <!-- Command Line Argument to Select Teleop Device -->
  <arg name="device" default="android" doc="available devices:
[keyboard, joystick, android]"/>

  <!-- Load the robot model into the parameter server -->
  <arg name="urdf_file" default="$(find xacro)/xacro --inorder
'$(find my_robot_model)/urdf/my_robo_$(arg mode).urdf'"/>
  <param name="robot_description" command="$(arg urdf_file)" />
```

```
<!-- Launch Hardware/Simulation Specific Modules -->
<include file="$(find bringup)/launch/mapping/$(arg mode).
launch.xml"/>

<!-- Joint State Publisher - Publishes Joint Positions -->
<node name="joint_state_publisher" pkg="joint_state_
publisher" type="joint_state_publisher"/>

<!-- Robot State Publisher  - Uses URDF and Joint States to
compute Transforms -->
<node name="robot_state_publisher" pkg="robot_state_
publisher" type="robot_state_publisher"/>

<!-- RVIZ  - Visualization -->
<node name="rviz" pkg="rviz" type="rviz" args="-d $(find my_
robot_model)/rviz/my_robo_mapping.rviz"/>

<!-- Select Teleop Device -->
<include file="$(find bringup)/launch/mapping/$(arg device).
launch.xml">
   <arg name="mode" value="$(arg mode)" />
</include>

<!-- SLAM - Map Building -->
<node name="gmapping" pkg="gmapping" type="slam_gmapping">
   <rosparam file="$(find bringup)/config/gmapping.yaml"
   command="load"/>
</node>

</launch>
```

This code shows the contents of the mapping.launch file. The launch file is written in XML. The important parts of the launch file are explained next:

```
<?xml version="1.0"?>
```

This line specifies the XML version. Here, it is 1.0.

```
<launch>
    ...
</launch>
```

This block represents the start and end of the launch file contents.

```
<arg name="mode" default="simulation" doc="available modes:
[hardware, simulation]"/>
```

This shows the command-line argument named mode, which indicates the mode of operation—simulation or hardware. This argument selects the appropriate launch file and URDF file.

```
<arg name="device" default="android" doc="available devices:
[keyboard, joystick, android]"/>
```

This shows the command-line argument named device, which indicates the device used to control the robot—keyboard, joystick, or android. This argument selects the appropriate launch file for teleop.

```
<arg name="urdf_file" default="$(find xacro)/xacro --inorder
'$(find my_robot_model)/urdf/my_robo_$(arg mode).urdf'"/>
<param name="robot_description" command="$(arg urdf_file)" />
```

Here, the robot description mentioned in the URDF is loaded into the parameter server. The parameter server stores the global data, which is accessible to all ROS nodes.

```
<include file="$(find bringup)/launch/mapping/$(arg mode).
launch.xml"/>
```

This runs another launch file, which is selected based on the mode command-line argument.

```
<node name="joint_state_publisher" pkg="joint_state_publisher"
type="joint_state_publisher"/>
```

Here, the joint state publisher node runs. The joint state publisher publishes the positions of all the moveable joints in the robot. In Bumblebot, the joint state publisher publishes the positions of left_wheel_joint, right_wheel_joint, front_castor_wheel_joint, and rear_castor_wheel_joint.

```
<node name="robot_state_publisher" pkg="robot_state_
publisher" type="robot_state_publisher"/>
```

Here, the robot state publisher node runs. The robot state publisher takes two inputs: the robot description from the URDF and the updated joint state values. Then, it computes forward kinematics and publishes the transforms. Forward kinematics calculates the new position and orientation based on the updated joint values of the robot.

```
<node name="rviz" pkg="rviz" type="rviz" args="-d $(find my_
robot_model)/rviz/my_robo_mapping.rviz"/>
```

Here, RViz runs and loads a previously saved RViz file called my_robo_mapping.rviz, which displays the robot model, generated map, Lidar scan, and so on.

```
<include file="$(find bringup)/launch/mapping/$(arg device).
launch.xml">
  <arg name="mode" value="$(arg mode)" />
</include>
```

This runs another launch file, which is responsible for reading inputs from the device and publishing velocity values for the robot. The launch file is selected based on the device command-line argument.

```
<node name="gmapping" pkg="gmapping" type="slam_gmapping">
  <rosparam file="$(find bringup)/config/gmapping.yaml"
  command="load"/>
</node>
```

This runs the slam_gmapping node, which is responsible for performing the mapping. A configuration file with several tuning parameters, called gmapping.yaml, is also loaded.

If everything goes well, you'll see the output shown in Figure 7-14 and the user will be able to perform mapping and teleoperation commands via the specified input device.

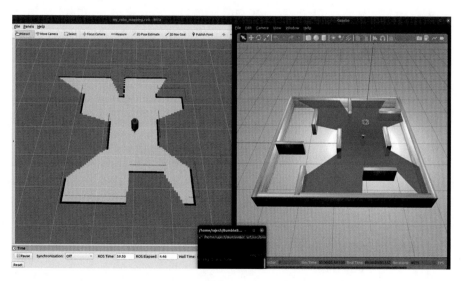

Figure 7-14. *Bumblebot performing mapping in simulation. RViz (left), Gazebo (right), and Keyboard teleop terminal (middle)*

You need to teleoperate the robot in the environment to generate the map. After the robot has explored the environment to the required extent and generated the map, you can save the map using this command:

```
rosrun map_server map_saver -f my_map
```

Here, the `map_saver` node is executed. It resides in the package named `map_server`. The `map_saver` node saves the generated map into the `my_map` file.

Navigation

Autonomous robot navigation is the ability of a robot to move from the initial position to the goal position without colliding with obstacles. Obstacles may be static (such as walls and partitions, which are immovable) or dynamic (such as people, chairs, and doors, which are subject to changes in position and can block the robot when in its vicinity).

Bumblebot uses map-based navigation. That is, the robot uses a previously built map of the environment for navigation. This enables the robot to plan an appropriate path toward the goal.

To simulate autonomous navigation in Bumblebot, type the following command in a terminal:

```
roslaunch bringup navigation_simulation.launch
```

Here, the `roslaunch` command launches the `bringup` package, which is the name of the ROS package where the launch file is located, and `navigation_simulation.launch` is the name of the launch file. A launch file in ROS starts multiple nodes or programs. Here are the contents of the `navigation.launch` file.

```
<?xml version="1.0"?>
<launch>

  <!-- Load the robot model into the parameter server -->
  <arg name="urdf_file" default="$(find xacro)/xacro --inorder
  '$(find my_robot_model)/urdf/my_robo_simulation.urdf'"/>
  <param name="robot_description" command="$(arg urdf_file)" />
```

```xml
<!-- Launch the gazebo world -->
<include file="$(find gazebo_ros)/launch/empty_world.launch">
  <arg name="world_name" value="$(find my_robot_model)/
  gazebo_worlds/plaza_world.world"/>
</include>

<!-- Load the robot model in the parameter server into the
gazebo world -->
<node name="urdf_spawner" pkg="gazebo_ros" type="spawn_model"
respawn="false" output="screen" args="-urdf -model my_robo
-param robot_description"/>

<!-- Joint State Publisher - Publishes Joint Positions -->
<node name="joint_state_publisher" pkg="joint_state_
publisher" type="joint_state_publisher"/>

<!-- Robot State Publisher  - Uses URDF and Joint States to
compute Transforms -->
<node name="robot_state_publisher" pkg="robot_state_
publisher" type="robot_state_publisher"/>

<!-- RVIZ  - Visualization -->
<node name="rviz" pkg="rviz" type="rviz" args="-d $(find my_
robot_model)/rviz/my_robo_navigation.rviz"/>

<!-- Map server -->
<node pkg="map_server" name="map_server" type="map_server"
args="'$(find my_robot_model)/maps/plaza_world_map.yaml'"/>

<!-- AMCL - Localization -->
<node pkg="amcl" type="amcl" name="amcl" output="screen">
    <rosparam file="$(find bringup)/config/amcl.yaml"
    command="load"/>
</node>
```

```
<!-- Move Base - Navigation -->
<node pkg="move_base" type="move_base" name="move_base"
 output="screen">
  <rosparam file="$(find bringup)/config/costmap_common_
  params.yaml" command="load" ns="global_costmap"/>
  <rosparam file="$(find bringup)/config/costmap_common_
  params.yaml" command="load" ns="local_costmap"/>
  <rosparam file="$(find bringup)/config/local_costmap_
  params.yaml" command="load" />
  <rosparam file="$(find bringup)/config/global_costmap_
  params.yaml" command="load" />
  <rosparam file="$(find bringup)/config/global_planner_
  params.yaml" command="load" />

  <!-- GLOBAL PLANNERS -->
  <!--param name="base_global_planner" value="navfn/
  NavfnROS" /-->
  <param name="base_global_planner" value="global_planner/
  GlobalPlanner"/>

  <!-- LOCAL PLANNERS -->
  <rosparam file="$(find bringup)/config/dwa_local_planner.
  yaml" command="load" />
  <param name="base_local_planner" value="dwa_local_planner/
  DWAPlannerROS"/>

  <!--rosparam file="$(find bringup)/config/trajectory_
  planner.yaml" command="load" />
  <param name="base_local_planner" value="base_local_planner/
  TrajectoryPlannerROS"/-->

  <!--rosparam file="$(find bringup)/config/teb_local_
  planner.yaml" command="load" />
```

```
  <param name="base_local_planner" value="teb_local_planner/
  TebLocalPlannerROS" /-->
 </node>
```

```
</launch>
```

This code shows the contents of the mapping.launch file. The launch file is written in XML. The important parts of the launch file are explained next:

```
<?xml version="1.0"?>
```

This line specifies the XML version. Here, it is 1.0.

```
<launch>
 ...
</launch>
```

This block represents the start and end of the launch file contents.

```
  <arg name="urdf_file" default="$(find xacro)/xacro --inorder
'$(find my_robot_model)/urdf/my_robo_simulation.urdf'"/>
  <param name="robot_description" command="$(arg urdf_file)" />
```

Here, the robot description mentioned in the URDF is loaded into the parameter server. The parameter server stores the global data, which is accessible to all ROS nodes.

```
<node name="urdf_spawner" pkg="gazebo_ros" type="spawn_model"
respawn="false" output="screen" args="-urdf -model my_robo
-param robot_description"/>
```

Here, the robot model in the parameter server is spawned into the Gazebo world.

```
<node name="joint_state_publisher" pkg="joint_state_publisher"
type="joint_state_publisher"/>
```

Here, the joint state publisher node runs. The joint state publisher publishes the positions of all the moveable joints in the robot. In Bumblebot, the joint state publisher publishes the positions of `left_wheel_joint`, `right_wheel_joint`, `front_castor_wheel_joint`, and `rear_castor_wheel_joint`.

```
<node name="robot_state_publisher" pkg="robot_state_
publisher" type="robot_state_publisher"/>
```

Here, the robot state publisher node runs. The robot state publisher takes two inputs: the robot description from the URDF and the updated joint state values. Then, it computes forward kinematics and publishes the transforms. Forward kinematics calculates the new position and orientation based on the updated joint values of the robot.

```
<node name="rviz" pkg="rviz" type="rviz" args="-d $(find my_
robot_model)/rviz/my_robo_navigation.rviz"/>
```

Here, RViz runs and loads a previously saved RViz file called `my_robo_navigation.rviz`, which displays the robot model, the map, the Lidar scan, the global path, the local path, and so on.

```
<node pkg="map_server" name="map_server" type="map_server"
args="'$(find my_robot_model)/maps/plaza_world_map.yaml'"/>
```

This runs the `map_server` node, which loads the specified map file. The map is necessary for the robot to perform autonomous navigation, because it provides information about static obstacles in the environment.

```
<node pkg="amcl" type="amcl" name="amcl" output="screen">
    <rosparam file="$(find bringup)/config/amcl.yaml"
    command="load"/>
</node>
```

This runs the localization node responsible for determining the position and orientation of the robot in the environment. A configuration file with several tunable parameters for localization—called amcl.yaml—is also loaded in this step.

```
<!-- Move Base - Navigation -->
<node pkg="move_base" type="move_base" name="move_base"
output="screen">
  <rosparam file="$(find bringup)/config/costmap_common_
  params.yaml" command="load" ns="global_costmap"/>
  <rosparam file="$(find bringup)/config/costmap_common_
  params.yaml" command="load" ns="local_costmap"/>
  <rosparam file="$(find bringup)/config/local_costmap_
  params.yaml" command="load" />
  <rosparam file="$(find bringup)/config/global_costmap_
  params.yaml" command="load" />
  <rosparam file="$(find bringup)/config/global_planner_
  params.yaml" command="load" />

  <!-- GLOBAL PLANNERS -->
  <!--param name="base_global_planner" value="navfn/
  NavfnROS" /-->
  <param name="base_global_planner" value="global_planner/
  GlobalPlanner"/>

  <!-- LOCAL PLANNERS -->
  <rosparam file="$(find bringup)/config/dwa_local_planner.
  yaml" command="load" />
  <param name="base_local_planner" value="dwa_local_planner/
  DWAPlannerROS"/>

  <!--rosparam file="$(find bringup)/config/trajectory_
  planner.yaml" command="load" />
  <param name="base_local_planner" value="base_local_planner/
  TrajectoryPlannerROS"/-->
```

```
<!--rosparam file="$(find bringup)/config/teb_local_
planner.yaml" command="load" />
<param name="base_local_planner" value="teb_local_planner/
TebLocalPlannerROS" /-->
</node>
```

This is the configuration for the move_base node used for robot navigation. The output parameter is set to screen, which means the output of running this node will be printed on the terminal.

The move_base node loads several parameter files (.yaml files) that define the costmap and planners for global and local navigation.

The base_global_planner parameter is set to global_planner/GlobalPlanner, which means the GlobalPlanner plugin will be used as the global planner. The global planner is responsible for generating a path from the robot's current position to its destination.

The base_local_planner parameter is set to dwa_local_planner/DWAPlannerROS, which means the DWAPlannerROS plugin will be used as the local planner. The local planner is responsible for generating velocity commands that follow the planned path while avoiding obstacles in real time.

There are also commented-out lines that show other local planners that can be used instead of DWAPlannerROS, such as TrajectoryPlannerROS and TebLocalPlannerROS. These lines are currently not active in the configuration.

If everything goes well, you'll see the output shown in Figures 7-15 and 7-16. The user will be able to provide a goal pose using the 2D Nav Goal button and click and drag on the map at the required location. This will command the robot to perform autonomous navigation.

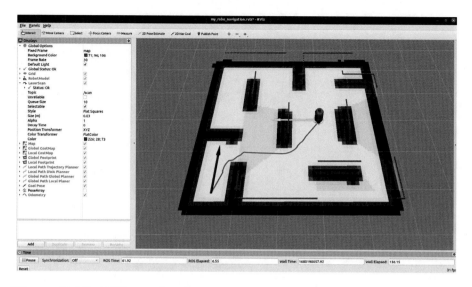

Figure 7-15. *RViz navigation*

In Figure 7-15, you can see the robot navigating toward the goal. The goal pose is indicated by the red arrow at the bottom-left corner of the map. The current robot pose is indicated by the yellow arrow in front of the robot. The blue line indicates the global path from the robot's current location to the target location. Note also the Lidar scan data represented in red.

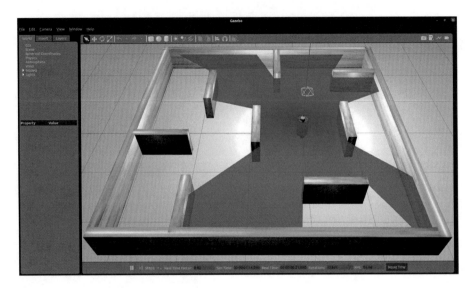

Figure 7-16. *Gazebo navigation*

Tuning Navigation

Tuning the navigation involves adjusting the parameters of the navigation stack to improve the performance of a robot's navigation system. Some of the frequently tuned parameters include:

- **Inflation radius**: This parameter determines how far the robot should stay away from obstacles. If the radius is too small, the robot might collide with obstacles, but if it is too large, the robot might take a longer path to reach its goal.

- **Path distance bias**: This controls the balance between following the planned path closely and deviating from the path to avoid obstacles or other hazards. When the path distance bias is set to a high value, the navigation stack will prioritize staying close to the planned path even if it means the robot might collide with an obstacle. On the other hand, when the path distance bias is set to a low value, the navigation stack will prioritize deviating from the planned path to avoid obstacles or other hazards, even if it means the robot might take a longer path to reach its goal.

- **Goal distance bias**: This controls the balance between finding a path that is the shortest distance to the goal and finding a safer path. When the goal distance bias is set to a high value, the navigation stack will prioritize shorter paths even if they are riskier, which means that the robot might take unsafe shortcuts to reach its goal. On the other hand, when the goal distance bias is set to a low value, the navigation stack will prioritize safer paths even if they are longer, which means that the robot might take a longer path to avoid obstacles and ensure safe navigation.

- **Global planner parameters**: These parameters control the behavior of the global planner, which plans a path from the robot's current position to the goal position. Typically tuned parameters include the planning algorithm, distance threshold, and goal tolerance to optimize the path-planning process.

- **Local planner parameters**: These parameters control the behavior of the local planner, which adjusts the robot's velocity to follow the planned path. Typically tuned parameters include planning frequency, maximum velocity, and acceleration limits to ensure smooth and safe navigation.

- **Recovery behavior parameters**: These parameters control the robot's behavior when it encounters unexpected obstacles or errors. Typically tuned parameters include the number of recovery attempts and the distance threshold for detecting obstacles to ensure that the robot can recover from unexpected situations.

To easily modify the parameters in real time and visualize the effects, you can use the `rqt_reconfigure` tool. You can start this tool using the following command:

```
rosrun rqt_reconfigure rqt_reconfigure
```

A screenshot of the tool is shown in Figure 7-17. You can see the parameters of the local planner (here, DWA) in the `move_base` node, which can be modified.

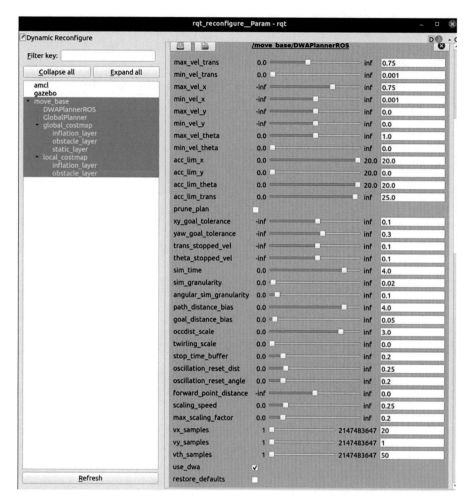

Figure 7-17. *The Rqt_Reconfigure tool*

You can adjust the parameters depending on your robot's hardware, environment, and navigation requirements. It's important to carefully evaluate the performance of your robot's navigation after adjusting any parameters to ensure that the robot can navigate safely and efficiently.

Summary

This chapter looked at the preliminary setup process, building the robot model (URDF), designing robot parts using 3D modeling software, adding 3D models inside URDF, robot visualization, robot simulation, teleoperation using the keyboard/joystick/Android devices, mapping, autonomous navigation, and tuning navigation. The next chapter explains how to configure Arduino to control peripherals, 3D printing robot parts, assembling and wiring the robot, and much more.

CHAPTER 8

Building Bumblebot in Hardware

Outline

This chapter discusses how to physically build a basic two-wheeled robot called "Bumblebot" using hardware components. It consists of the following steps:

- Preliminary robot setup

- Udev rules

- Configuring Arduino to control peripherals

- 3D printing the robot parts

- Assembling and wiring the electronics

- Motor gear ratio calculation

- Custom motor driver and ROS interface

- Differential driver and odometry

- Teleoperation using ROS

- Odometry correction: rotation and translation

© Rajesh Subramanian 2023
R. Subramanian, *Build Autonomous Mobile Robot from Scratch using ROS*,
Maker Innovations Series, https://doi.org/10.1007/978-1-4842-9645-5_8

- Map building

- Autonomous navigation

- Navigation tuning

- Robot upstart

- Autonomous delivery application

The Preliminary Robot Setup Process

This setup involves configuring Odroid XU4, which is the single-board computer (SBC) used in this robot. It is similar to Raspberry Pi and is capable of running various Ubuntu distributions with the Robot Operating System (ROS). Bumblebot uses Ubuntu 20.04 (Focal Fossa) with ROS Noetic.

To configure the robot, you can either flash a pre-built image of it into a microSD card or you can set it up from scratch.

Method 1: Using a pre-built robot image:

1. Download the image.

2. You need to download the image file with all the required software and configuration files for the robot's operation: `https://drive.google.com/file/d/19kr7EIDr1K4oltk4DdJyzPR-TPoS2oON/view?usp=share_link`

3. Unzip the downloaded file.

4. Flash the image into a microSD card using software such as Etcher (in Linux), Rufus (in Windows), or any similar software.

Method 2: Setting it up from scratch:

1. Download the Ubuntu Mate image for Odroid from this link: `https://odroid.in/ubuntu_20.04lts/XU3_XU4_MC1_HC1_HC2/ubuntu-20.04-5.4-mate-odroid-xu4-20210926.img.xz`

2. Flash the image into a microSD using software such as Etcher (in Linux), Rufus (in Windows), or any similar software.

3. Install ROS Noetic by following the instructions given in the following link: `http://wiki.ros.org/noetic/Installation/Ubuntu`.

4. Clone the Bumblebot repository from GitHub using the following command:

    `git clone https://github.com/logicraju/BumbleBot_WS.git`

5. Go to the robot workspace using the following command:
    `cd BumbleBot_WS`

6. In Git, you can create multiple branches (versions) of the software. Here, the default branch is ROS Noetic. If you're using ROS Noetic, skip this step. If you're using ROS Melodic, change the Git branch to `melodic-devel` using the following command:

    `git checkout origin/melodic-devel`

7. Install the required ROS packages and dependencies using the following command:

    `rosdep install --from-paths src --ignore-src -r -y`

8. Compile the workspace using the following command:

```
cd ~/Bumblebot_WS
catkin_make
```

9. Source the workspace using the following command:

```
source ~/Bumblebot_WS/devel/setup.bash
```

To permanently source the workspace, add the previous command to the end of the ~/.bashrc file, as shown in Figure 8-1.

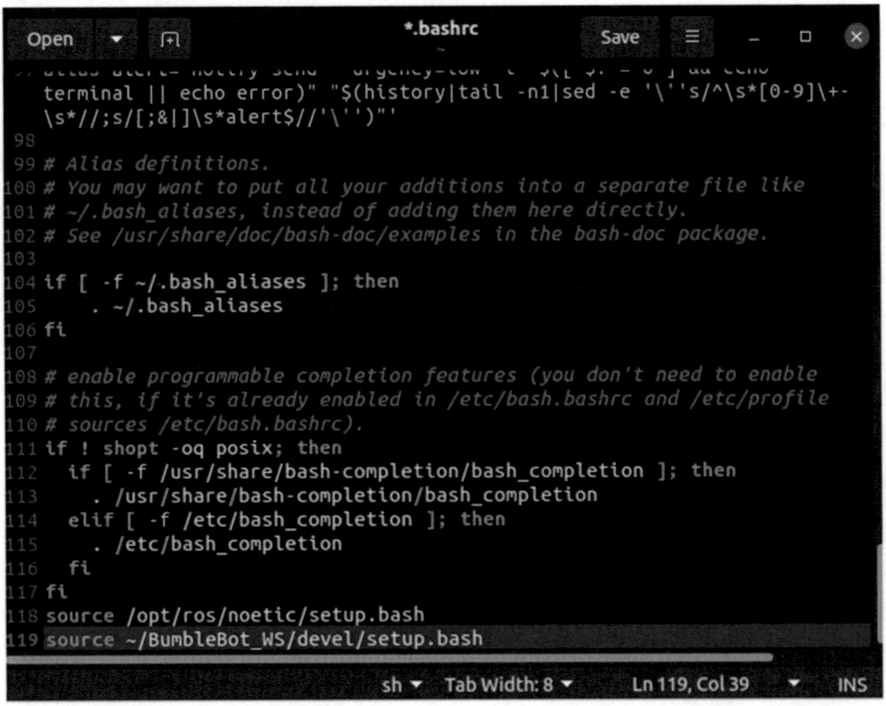

Figure 8-1. *The .bashrc file*

10. If you are using multiple machines, (here, Android and SBC), you need to configure the network setup of the robot to establish the necessary communication. For this, you need to specify certain environment variables to ROS. A template is provided in this file:

BumbleBot_WS/src/resources/Bashrc/Robot/.bashrc

Copy the following lines from the template file into the ~/.bashrc file of the robot:

```
#ROBOT NETWORK CONFIGURATION
robot_ip=$(ip route get 8.8.8.8 | awk -F"src "
'NR==1{split($2,a," ");print a[1]}')
export ROS_MASTER_URI=http://localhost:11311
export ROS_HOSTNAME=$robot_ip
export ROS_IP=$robot_ip
echo "ROBOT"
echo "ROS_HOSTNAME: "$ROS_HOSTNAME
echo "ROS_IP: "$ROS_IP
echo "ROS_MASTER_URI: "$ROS_MASTER_URI
```

The robot_ip variable contains the IP address of the robot and the variable will be updated with the current IP address of the robot. This is useful if the robot is assigned a dynamic IP address. You can also assign a static IP to the robot.

Now, you specify the environment variables for the ROS network setup of the robot:

- ROS_MASTER_URI signifies where the ROS master node is running. It is specified as localhost because the ROS master will be running in the robot. The number 11311 is the default port number.

- ROS_IP specifies the IP address of the robot.

- ROS_HOSTNAME specifies the hostname of the robot. You can also specify the IP address instead of the hostname. In this case, the IP address of the robot is specified.

Udev Rules

A Udev rule is a configuration file that defines how devices should be named, identified, and managed on a Linux system. Udev rules are written in a simple scripting language and are stored in the /etc/udev/rules.d/ directory on most Linux systems. The filename of a Udev rule must end with the .rules extension. When a device is connected or disconnected from the system, Udev reads the Udev rules and performs the actions defined in them.

A typical Udev rule consists of a set of conditions that must be met for the rule to be applied and a set of actions that Udev should perform when the rule is applied. The conditions can include attributes of the device, such as its vendor and product IDs, its device type, or its physical location on the system. The actions can include renaming the device node, creating a symlink to the device node, or running a script to configure the device. For example, a Udev rule can be used to automatically mount a USB drive when it is connected to the system or to assign a consistent name to a network interface regardless of its physical location.

Defining Rules

You can use Udev rules to permanently set the labels and the read/write/ execute permissions for a hardware device that's connected to the robot. These unique names and permissions enable various ROS nodes to access the hardware devices of the robot. Follow these steps:

1. Connect the device to the USB port of the computer.
    This will generate a device file in the /dev folder.

2. Open a terminal.

3. Navigate to the /dev folder using this command:

    cd /dev

4. List the device names using this command (see
    Figure 8-2):

    ls -l | grep -i -E "tty[a-z]"

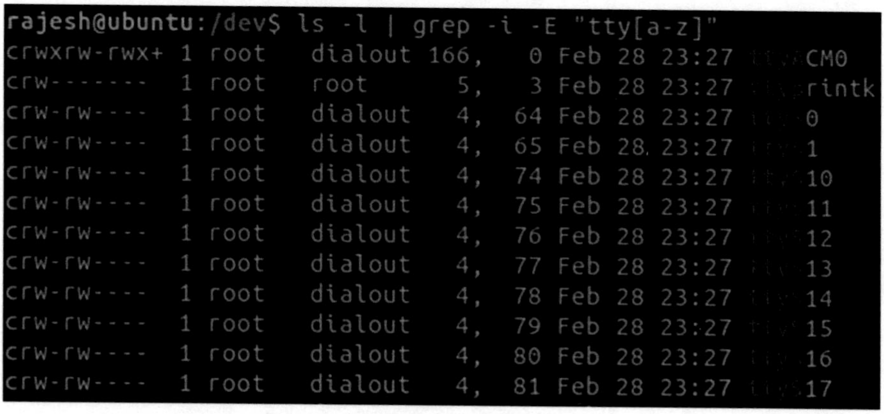

```
rajesh@ubuntu:/dev$ ls -l | grep -i -E "tty[a-z]"
crwxrw-rwx+ 1 root    dialout 166,   0 Feb 28 23:27      CM0
crw------- 1 root    root      5,   3 Feb 28 23:27      rintk
crw-rw---- 1 root    dialout   4,  64 Feb 28 23:27      0
crw-rw---- 1 root    dialout   4,  65 Feb 28. 23:27     1
crw-rw---- 1 root    dialout   4,  74 Feb 28 23:27      10
crw-rw---- 1 root    dialout   4,  75 Feb 28 23:27      11
crw-rw---- 1 root    dialout   4,  76 Feb 28 23:27      12
crw-rw---- 1 root    dialout   4,  77 Feb 28 23:27      13
crw-rw---- 1 root    dialout   4,  78 Feb 28 23:27      14
crw-rw---- 1 root    dialout   4,  79 Feb 28 23:27      15
crw-rw---- 1 root    dialout   4,  80 Feb 28 23:27      16
crw-rw---- 1 root    dialout   4,  81 Feb 28 23:27      17
```

Figure 8-2. *List of device names*

5. Identify the device name (such as ttyACM0). This
    can be done by disconnecting and reconnecting the
    device to the computer and listing the devices. The
    device name is the one that disappears when you're
    disconnected and reappears you're reconnected.

6. Get the device information by typing the following
    command:

    udevadm info /dev/ttyACM0 | grep -E 'ID_
    VENDOR_ID|ID_MODEL_ID|ID_SERIAL_SHORT'

Where `ttyACM0` is the name of the device file of the device. This command will display the vendor ID, model ID and serial number, as shown in Figure 8-3.

```
rajesh@ubuntu:/dev$ udevadm info /dev/ttyACM0 | grep -E 'ID_VENDOR_ID|ID_MODEL_I
D|ID_SERIAL_SHORT'
E:           =2341
E:           =0043
E:           =75630313636351A052C1
```

Figure 8-3. *Device information displayed using the udevadm command*

This device information allows you to identify the device each time it is plugged into the computer.

7. Next, you need to create a rules file with the device information and assign a label to the device of your choice. To do this, type the following commands:

```
cd /etc/udev/rules.d/
sudo gedit my_rules.rules
```

Now type the following details into the document and save it:

```
SUBSYSTEM=="tty", SUBSYSTEMS=="usb",
ATTRS{idVendor}=="2341",
ATTRS{idProduct}=="0043",
ATTRS{serial}=="75630313636351A052C1",
MODE="0777", SYMLINK+="arduino_uno"
```

Next, assign values to the `idVendor` and `idProduct` variables and get serial number from the device information you retrieved in Step 6. Here, `MODE=0777` assigns read/write and execute permissions to the device. Also, `SYMLINK+="arduino_uno"` assigns a custom label to the device (here, `arduino_uno` is the custom label).

8. Next, type the following command to make the rules take effect:

```
sudo udevadm control --reload-rules && sudo
service udev restart && sudo udevadm trigger
```

9. To verify whether the rules have taken effect, you can type the following commands:

```
cd /dev
ls -l | grep arduino
```

If everything went well, you'll see the output in Figure 8-4.

```
rajesh@ubuntu:~$ cd /dev
rajesh@ubuntu:/dev$ ls -l | grep arduino
lrwxrwxrwx  1 root    root           7 Mar  1 08:12        _uno -> ttyACM0
```

Figure 8-4. *Output of the command*

Here, you can see that the device is assigned the name `arduino_uno`. Also, `rwx` on the left indicates that read/write/execute permissions are assigned. From now on, whenever you unplug and plug the device back in, the name along and its permissions are reassigned based on your rules.

A template Udev rule file for reference is provided in the robot workspace at `BumbleBot_WS/src/resources/Udev Rules/my_robot.rules`.

Configuring Arduino to Control Peripherals

Arduino Mega controls the status LEDs, the LCD screen, the pushbutton, and the buzzer of the robot. To set up Arduino, follow these steps:

1. Start the Arduino IDE.

2. Open the code provided at:

```
BumbleBot_WS/src/arduino_codes/codes/
LED_LCD_Buzzer_Switch_Protothread/LED_LCD_
Buzzer_Switch_Protothread.ino
```

3.   Flash the code into Arduino Mega using the IDE.

The Arduino code includes ROS publishers and subscribers to communicate with the robot's computer (Odroid XU4). The Arduino code does the following:

1.   Reads the pushbutton status and publishes under the /push_button topic.

2.   Listens to messages under the /led topic and triggers status LEDs appropriately.

3.   Listens to messages under the /lcd topic and displays the message on the LCD screen.

4.   Listens to messages under the /buzzer topic and plays music notes on the buzzer.

3D Printing the Robot Parts

To 3D print the parts in this example, a good, budget printer called Creality Ender 3 V2 was used. All the robot parts were printed using PLA Pro+ filament material. If a 3D printer or a 3D printing facility is not available, you can assemble the components in a suitable box or container and modify the robot kinematic description in the URDF file. Before 3D printing the 3D models, you need to convert the models from the .stl (or any equivalent) format into layers in gcode format. This is because a 3D printer prints layer by layer. There are various slicing tools available, including Ultimaker Cura, which is free, widely used, and provides numerous customizable settings. The slicing tool can be downloaded from the following website:

https://ultimaker.com/software/ultimaker-cura

The settings used to print the robot parts for Bumblebot are shown in Figure 8-5.

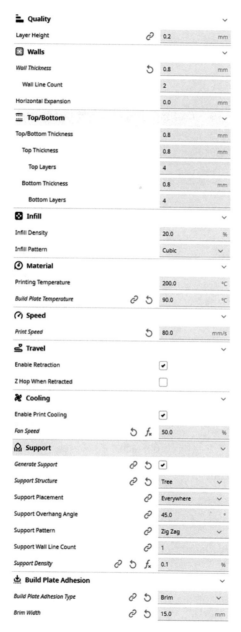

Figure 8-5. *Cura settings*

Electronics

The complete list of the electronic components used for Bumblebot is given in Table 8-1.

Table 8-1. *List of Electronic Components Used in Bumblebot*

#	Component	Purpose
1	Odroid XU4	Runs the required software for the robot
2	Samsung EVO Plus 32 GB MicroSD card	Installs Ubuntu, ROS, and other software
3	Arduino Mega 2560	Communicates with ROS, controls the LCD, buzzer, LEDs, and pushbutton
4	RMCS2303 Motor Driver	Controls the speed of motors, gets feedback, applies the PID
5	22.2V Li-Ion Battery	Provides power to the robot
6	Step Down Circuits	Converts the battery voltage to the required voltages (12V, 5V)
7	Lidar A1M8	Gets the distance between the robot and any obstacles
8	DC Motor with Encoder RMCS5025	Turns the wheels and provides motor position
9	Powered USB Hub 4 Port	Connects various USB devices to Odroid XU4
10	IMax B6 5A Battery Charger	Recharges the battery
11	CP2102 USB to TTL	Connects the motor controller to Odroid XU4
12	USB Extender Cable	Connects external devices to the robot after assembly

(continued)

Table 8-1. (*continued*)

#	Component	Purpose
13	TP Link WN725N WiFi Adapter	Connects the robot to a WiFi network
14	Redgear Pro Series Gamepad	Controls the robot manually and triggers various functionalities
15	Wireless Keyboard and Mouse	Interacts with the robot
16	Micro USB (type B) to USB (type A) cable	Connects RPLidar to Odroid XU4
17	Rocker Switch	Toggles the power to the robot
18	LCD – 16X2	Displays status messages
19	I2C Serial Adapter	Converts I2C serial data to parallel data for LCD
20	Battery Capacity Monitor	Monitors battery voltage
21	LED	Provides status indications
22	Pushbutton – Two Pin	Interacts with the robot
23	HDMI Dummy Plug	Runs OdroidXU4 in headless mode and displays RViz
24	330 Ohm Resistors	Limits current drawn by the LED
25	10KOhm Resistors	Prevents the pushbutton output from varying
26	DC Power Supply SMPS 24V, 5A	Tests the robot without a battery

Assembling and Wiring

After 3D printing the robot's parts and getting the required electronic components, you need to perform the assembling and wiring processes. The components and tools you need to assemble Bumblebot are as follows:

- 3D-printed robot parts
- Electronic components
- Castor wheels (x2)
- Wheels (x2)
- Standard nuts, self-clinch nuts/rivet nuts, and bolts (3mm diameter)
- Screwdriver
- Wrench
- Nose plier
- Multimeter
- Multi-strand wires (1mm2)
- Stand-offs (3mm diameter)
- Glue gun and glue sticks
- Soldering iron and stand
- Solder wire
- Solder flux
- Desoldering pump

Wiring

Odroid XU4 is a single-board computer that runs ROS and other necessary software. It is connected to other electronic devices via USB and HDMI ports. A powered USB port connects Lidar, as Odroid XU4 does not supply the required power. The USB hub also provides additional USB ports to connect the devices.

Arduino Mega is connected to Odroid XU4 via a USB port. Arduino Mega controls the robot's status and interaction devices (such as LEDs, LCD, pushbuttons, and the buzzer). Power is provided to the motors and USB hub via a 12V step-down circuit. Arduino Mega is powered by the USB port of Odroid XU4. Odroid XU4 is powered by a 5V step-down circuit. The wiring diagram of Bumblebot is depicted in Figure 8-6.

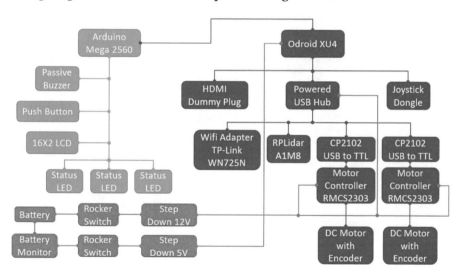

Figure 8-6. *Wiring diagram for Bumblebot*

The connection diagram of the motor controller (RMCS2303) with the servo motor (RMCS5025) is depicted in Figure 8-7. The USB-to-TTL converter (CP2102) connects the motor controller to Odroid XU4.

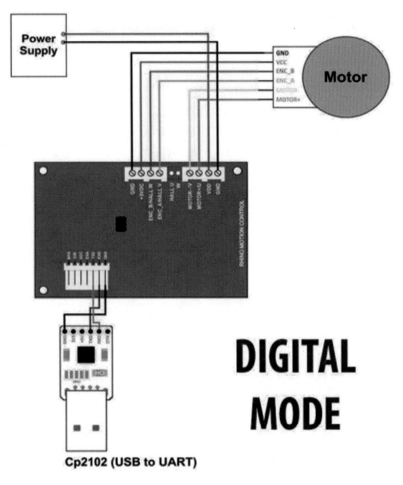

Figure 8-7. Wiring diagram of the motor controller and motor

The connection diagram of Arduino Mega with LEDs, LCD, pushbutton, and buzzer is shown in Figure 8-8.

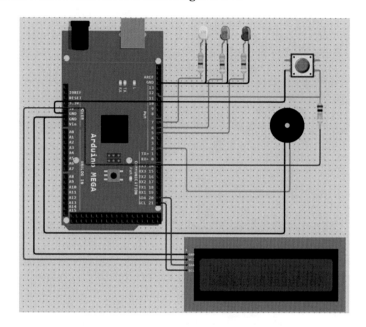

Figure 8-8. *Arduino Mega connection diagram*

Assembling the Robot

All the 3D model files of the robot are available in the robot workspace in the following path: BumbleBot_WS/src/resources/3D Print STLs/.

These files can be used for 3D printing after converting into gcode using suitable slicing software.

Figure 8-9 illustrates the 3D printed parts of Bumblebot before they have been assembled.

Figure 8-9. *3D printed parts of the robot before assembling (Note: some parts were redesigned later)*

Note   Some parts of the robot were redesigned and used in the final assembly.

Figure 8-10 shows the assembled electronics and 3D printed parts before mounting the outer cover.

Figure 8-10. *Bumblebot inner assembly*

Figure 8-11 depicts the robot after complete assembly and wiring.

Figure 8-11. *Bumblebot after complete assembly*

Motor Gear Ratio Calculation

The actual gear ratio of a motor can differ from that given in the manual/ datasheet. To command the motors to rotate at a specific speed, you need to specify the speed value in rotations per minute (RPM). But the velocity commands issued by the navigation stack are in radians per second. To convert radians per second to RPM, you need to find the gear ratio of the motors. Follow these steps to find the gear ratio:

1. In the servo motor (RMCS5025) used in Bumblebot, there will be 334 encoder pulses generated per complete rotation.

2. Rotate the motor for a specific number of times (say, 20 rotations).

3. Note the motor position after 20 rotations.

4. In Bumblebot, the motor position after 20 rotations is 1464404.

Note   This value is obtained from the Rhino motor software.

---

5. As the servo motor uses a quadrature encoder, it will generate four times the number of pulses per rotation.

   Gear ratio = motor position/(4*pulses per rotation * number of rotations)

6. Therefore, the gear ratio equals 1464404/(4*334*20) = 52.195 (60 on the datasheet).

To convert the velocity commands generated by the ROS navigation stack from radians per second to RPM, use this formula:

$$RPM = \text{radians per sec} * 9.549297 * \text{gear ratio}$$

If the navigation stack produces a velocity of one radian per second, you can command the motors to rotate at a speed of 1*9.549297*52.195, or 498.426 RPM.

The configuration and testing software for the motor controller can monitor the position of the motor while it is rotating. The software can be obtained from the following link:

https://robokits.co.in/downloads/Rhino%20DC%20Servo%20 2303%20Config%20Setup.exe

The following link shows you how to command the motor and obtain the position/speed feedback using this software.

https://www.youtube.com/watch?v=13Na9REGJUY&ab_ channel=MakeItIdeas

A sample screen of the software while testing the motor is shown in Figure 8-12.

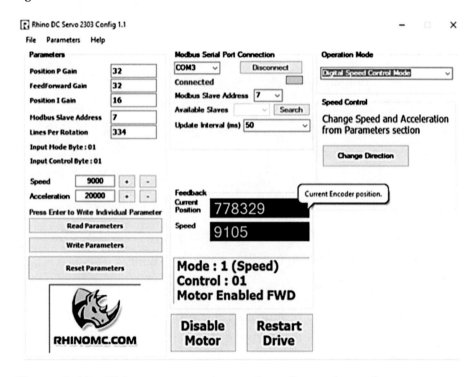

Figure 8-12. *Rhino motor testing and configuration software screenshot*

Custom Motor Driver and ROS Interface

For the robot to communicate with the motor driver circuit (RMCS2303), a custom motor driver was written in Python for this project. The driver code uses the Modbus protocol to communicate with the motor controller circuit. The driver code has these basic functionalities:

- Set speed

- Set acceleration

- Turn motor clockwise

- Turn the motor counter-clockwise

- Stop motor

- Get position

- Get speed

- Brake

The code can be found at `BumbleBot_WS/src/my_motor_controller/ src/motor_driver.py`.

On top of the motor driver code, interface code was also added to communicate with the other ROS nodes. This program publishes and subscribes to ROS topics and communicates with the motor driver code accordingly. The main functions the ROS interface node takes care of are as follows:

- Subscribes to the motor speed commands

- Scales the speed

- Communicates with the motor driver to set the motor speeds

- Gets the encoder ticks from the motor driver

- Publishes the encoder values

The code can be at `BumbleBot_WS/src/my_motor_controller/src/motor_driver_ros_interface.py`.

Differential Driver and Odometry

Differential drive configuration is a drive system with two driving wheels that turn independently and guide the robot. A ROS package called `differential_drive` written by Jon Stefan is used in this project.

- The `differential_drive` package contains ROS nodes to receive the velocities sent out by the ROS navigation stack.

- It then computes the individual motor velocities in radians per second and publishes them.

- The individual motor velocities are captured by the custom motor interface node (mentioned in the previous section).

- The custom motor interface node then commands the motor driver to operate the motors accordingly.

The `differential_drive` package can be found at `http://wiki.ros.org/differential_drive`.

Odometry involves estimating the robot's current location with respect to the starting position. This estimate is determined with the help of signals obtained from the encoders attached to the motors. The Odometry node publishes the current position, orientation, and velocity of the robot. Odometry data is used by the localizing node (here, the AMCL node) to localize the robot pose in the map. Odometry data is also used by the local planner to compute the next velocity to guide the robot toward the goal pose eventually.

The `differential_drive` package has two nodes:

- `twist_to_motors.py`: Differential drive node
- `diff_tf.py`: Odometry node

These nodes can be found at `BumbleBot_WS/src/differential_driver/src`.

twist_to_motors.py

The `twist_to_motors.py` code is a ROS node that splits the velocities given by the local planner into left and right wheel velocities.

In this code, you need to specify the base width of the robot in meters:

```
self.w = rospy.get_param("~base_width", 0.235)
```

You should specify the topic names of left and right wheel velocities, as follows:

```
self.pub_lmotor = rospy.Publisher('lwheel_vtarget', Float32,
queue_size=1)
self.pub_rmotor = rospy.Publisher('rwheel_vtarget', Float32,
queue_size=1)
```

Also, you need to add the topic name under which the velocity commands are published from the local planner:

```
rospy.Subscriber('/diffbot_controller/cmd_vel', Twist, self.
twistCallback)
```

The following are the equations used to compute left and right wheel velocities from the robot velocity:

```
self.right = 1.0 * self.dx - self.dr * self.w / 2
self.left = -1.0 * self.dx - self.dr * self.w / 2
```

diff_tf.py

The `diff_tf.py` code is a ROS node that computes odometry and transformations from the left and right encoder ticks.

You need to specify the base width of the robot in meters:

```
self.base_width = float(rospy.get_param('~base_width', 0.235))
```

You also need to mention the encoder ticks per meter in the code, as shown:

```
self.ticks_meter = float(rospy.get_param('ticks_meter',
279118))
```

To get the value of ticks per meter, you need to teleop the robot by one meter along a straight line and note the left and right encoder values. Encoder values can be obtained by subscribing to encoder topics. In Bumblebot, you can subscribe to the left and right encoder topics by using the following commands:

```
rostopic echo /lwheel_ticks_32_bit
rostopic echo /rwheel_ticks_32_bit
```

You can either take the ticks of the left encoder or the right encoder. For better accuracy, you can take the average value of the two encoder values, obtained after the robot traverses one meter, to get "ticks per meter."

You should also provide the frame name of the base footprint and odometry:

```
self.base_frame_id = rospy.get_param('~base_frame_id', 'base_
footprint')
self.odom_frame_id = rospy.get_param('~odom_frame_id', 'odom')
```

Then, you need to specify the minimum and maximum values of the encoders:

```
self.encoder_min = rospy.get_param('encoder_min', 0)
self.encoder_max = rospy.get_param('encoder_max', 4294967295)
```

The minimum and maximum encoder values can be obtained from the device manual.

Teleoperation

Teleoperation involves controlling the movements of the robot using an input device such as a keyboard, joystick, mobile phone, and so on. This includes the following:

- Reading the input keys and computing the linear and rotational velocities depending on the key pressed.

- Publishing the velocities (for the robot as a whole).

- Subscribing the velocities.

- Splitting the velocities into left/right wheel velocities.

- Computing the motor speeds.

- Generating corresponding voltages to turn the motors.

The next section explains how to teleoperate the Bumblebot using a keyboard, joystick, and Android phone.

Teleoperation Using Keyboard

To teleoperate Bumblebot using a keyboard, open a terminal and enter the following command:

```
roslaunch bringup teleop.launch device:=keyboard mode:=hardware
```

Here, the roslaunch command launches a bringup package, which is the name of the ROS package in which the launch file is located. teleop.

launch is the name of the launch file. A launch file in ROS starts multiple nodes or programs. The device:=keyboard and mode:=hardware values are command-line arguments. These arguments allow you to run the required launch files and nodes while executing the roslaunch command. Take a look at the contents of the teleop.launch file.

```xml
<?xml version="1.0"?>
<launch>

  <!-- Command Line Argument to Select Mode -->
  <arg name="mode" default="simulation" doc="available modes:
  [hardware, simulation]"/>

  <!-- Command Line Argument to Select Teleop Device -->
  <arg name="device" default="android" doc="available devices:
  [keyboard, joystick, android]"/>

  <!-- Load the robot model into the parameter server -->
  <arg name="urdf_file" default="$(find xacro)/xacro --inorder
  '$(find my_robot_model)/urdf/my_robo_$(arg mode).urdf'"/>
  <param name="robot_description" command="$(arg urdf_file)" />

  <!-- Launch Hardware/Simulation Specific Modules -->
  <include file="$(find bringup)/launch/teleop/$(arg mode).
  launch.xml"/>

  <!-- Joint State Publisher - Publishes Joint Positions -->
  <node name="joint_state_publisher" pkg="joint_state_
  publisher" type="joint_state_publisher"/>

  <!-- Robot State Publisher  - Uses URDF and Joint States to
  compute Transforms -->
  <node name="robot_state_publisher" pkg="robot_state_
  publisher" type="robot_state_publisher"/>
```

```
<!-- RVIZ  - Visualization -->
<node name="rviz" pkg="rviz" type="rviz" args="-d $(find my_
robot_model)/rviz/my_robo_teleop.rviz"/>

<!-- Select Teleop Device -->
<include file="$(find bringup)/launch/teleop/$(arg device).
launch.xml">
  <arg name="mode" value="$(arg mode)" />
</include>
```
```
</launch>
```

The previous code shows the contents of the teleop.launch file. A launch file is written in XML. The important parts of the launch file are explained next:

```
<?xml version="1.0"?>
```

This line specifies the XML version. Here, it is 1.0.

```
<launch>
  ...
</launch>
```

This block represents the start and end of the launch file contents.

```
<arg name="mode" default="simulation" doc="available modes:
[hardware, simulation]"/>
```

The command-line argument named mode indicates the mode of operation—simulation or hardware. This argument selects the appropriate launch file and URDF file.

```
<arg name="device" default="android" doc="available devices:
[keyboard, joystick, android]"/>
```

The command-line argument named `device` indicates the device used to control the robot—`keyboard`, `joystick`, or `android`. This argument selects the appropriate launch file for teleop.

```
<arg name="urdf_file" default="$(find xacro)/xacro --inorder
'$(find my_robot_model)/urdf/my_robo_$(arg mode).urdf'"/>
<param name="robot_description" command="$(arg urdf_file)" />
```

Here, the robot description mentioned in the URDF is loaded into the parameter server. The parameter server stores the global data, which is accessible to all ROS nodes.

```
<include file="$(find bringup)/launch/teleop/$(arg mode).
launch.xml"/>
```

This runs another launch file, which is selected based on the `mode` command-line argument.

```
<node name="joint_state_publisher" pkg="joint_state_publisher"
type="joint_state_publisher"/>
```

Here, the joint state publisher node runs. The joint state publisher publishes the positions of all the moveable joints in the robot. In Bumblebot, the joint state publisher publishes the positions of `left_wheel_joint`, `right_wheel_joint`, `front_castor_wheel_joint`, and `rear_castor_wheel_joint`.

```
<node name="robot_state_publisher" pkg="robot_state_
publisher" type="robot_state_publisher"/>
```

Here, the robot state publisher node runs. The robot state publisher takes two inputs: the robot description from the URDF and the updated joint state values. Then, it computes forward kinematics and publishes the transforms. Forward kinematics calculates the new position and orientation based on the updated joint values of the robot.

```
<node name="rviz" pkg="rviz" type="rviz" args="-d $(find my_
robot_model)/rviz/my_robo_teleop.rviz"/>
```

Here, RViz runs and loads a previously saved file named my_robo_
teleop.rviz, which displays the robot model, lidar scan, and so on.

```
<include file="$(find bringup)/launch/teleop/$(arg device).
launch.xml">
  <arg name="mode" value="$(arg mode)" />
</include>
```

This runs another launch file, which reads data from the input device
and publishes velocity values to the robot. The launch file is selected based
on the device command-line argument.

This launch file also starts another launch file, which is hardware-
specific and named hardware.launch.xml, as shown here:

```
<?xml version="1.0"?>
<launch>
  <!-- RPLidar A1M8 -->
  <include file="$(find rplidar_ros)/launch/rplidar.launch"/>

  <!-- Motor Controller -->
  <node pkg="my_motor_controller" type="motor_driver_ros_
interface.py" name="motor_controller_ros" output="screen"/>

  <!-- Differential Drive Controller  -->
  <node pkg="differential_driver" type="twist_to_motors.py"
name="twist_to_motors" output="screen"/>
  <!--node pkg="differential_driver" type="diff_tf.py"
name="diff_tf" output="screen"/-->
```

```
<!-- Arduino -->
<node pkg="rosserial_python" name="rosserial_arduino_node"
type="serial_node.py" args="/dev/arduino_mega" />
```

```
</launch>
```

The launch file includes several nodes that will run when the application starts.

- rplidar.launch launches the ROS driver for the RPLidar A1M8 sensor. This node will publish Lidar data on the ROS network.

- motor_driver_ros_interface.py is a custom node that interfaces with a motor controller. This node will subscribe to ROS topics and publish commands to control the motors and get encoder feedback.

- twist_to_motors.py is a ROS node that splits the velocities given by the local planner into left and right wheel velocities.

- rosserial_arduino_node is a ROS node that interfaces with an Arduino board using the rosserial library. This node will publish and subscribe to ROS topics to communicate with the Arduino board.

If everything goes well, the user will be able to issue commands via keyboard keys and control the robot. The keyboard commands must be issued while the cursor is in the terminal where the launch file is run.

You can also specify joystick or android as the command-line argument instead of keyboard to control the robot using the desired device.

Odometry Correction

Odometry requires accurately measuring the rotation of the wheels, but errors can accumulate over time. The error sources include wheel slippage, changes in wheel diameter, uneven terrain, and so on. As a result, the robot's estimate of its position can drift over time, leading to errors in its trajectory. There are two tests used to check if the odometry is reliable:

- Rotation test

- Translation test

Rotation Test

In the rotation test, you determine how much the odometry data drifts while rotating the robot. You do this by visualizing the laser scan data in RViz while rotating the robot. Ideally, the scan lines will fall directly on top of each other. If there is an error in the odometry data, the laser scan will drift further when the robot rotates. The laser scan can be visualized using RViz by enabling the Laser Scan tool. To perform the rotation test, do the following:

1. Type any one of the following commands in a terminal:

    ```
    roslaunch bringup odom_check.launch
    mode:=hardware device:=keyboard
    roslaunch bringup odom_check.launch
    mode:=hardware device:=joystick
    roslaunch bringup odom_check.launch
    mode:=hardware device:=android
    ```

2. This will enable the robot to be teleoperated using the specified input device and also displays the RViz window, as shown in Figure 8-13.

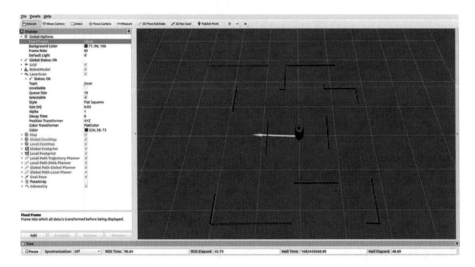

Figure 8-13. *RViz visualization while odometry correction test*

3.   Ensure that Fixed Frame under the Global Options
     menu in RViz is set to odom, as shown in Figure 8-14.

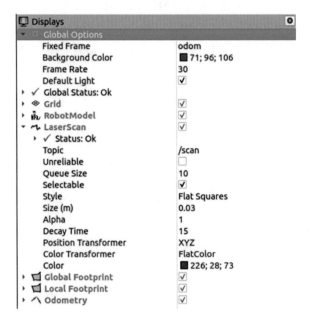

Figure 8-14. *RViz configuration for the odometry test*

4. Confirm that the Decay Time under the LaserScan menu is set to a high value, say 15 seconds, as shown in Figure 8-14.

5. Rotate the robot 360 degrees.

6. Check the angular drift between the initial position and final position of laser scan lines.

7. If the difference between the scan lines is below ten degrees, the odometry error is within acceptable limits.

8. Otherwise, you need to correct odometry by using one the following methods:

   a. Calibrating the sensor properly

   b. Using odometry correcting algorithms such as AMCL

   c. Replacing the sensor with a more accurate one

   d. Adding sensors like IMU, visual odometry, and so on

Translation Test

In the translation test, you determine how much the odometry data drifts while moving the robot along a straight line. You do this by visualizing the laser scan data in RViz while moving the robot. Ideally, the scans will fall directly on top of each other. If there is an error in the odometry data, the laser scan lines will drift further while the robot is moving. The laser scan can be visualized using RViz by enabling the Laser Scan tool. To perform the translation test, follow these steps:

1. Type any one of the following commands in a terminal:

    ```
    roslaunch bringup odom_check.launch
    mode:=hardware device:=keyboard
    roslaunch bringup odom_check.launch
    mode:=hardware device:=joystick
    roslaunch bringup odom_check.launch
    mode:=hardware device:=android
    ```

    This will enable the robot to be teleoperated using the specified input device and will also display the RViz window, as shown in Figure 8-13.

2. Ensure that Fixed Frame under the Global Options menu in RViz is set to odom, as shown in Figure 8-14.

3. Confirm that the Decay Time under the LaserScan menu is set to a high value, such as 15 seconds, as shown in Figure 8-14.

4. Move the robot forward by about one meter.

5. Check the linear drift between the initial and final positions of the laser scan lines.

6. If the difference between the scan lines is below about three centimeters, the odometry error is within acceptable limits.

7. Otherwise, you need to correct the odometry by using one of the following methods:

    •   Calibrating the sensor properly

    •   Using odometry correcting algorithms such as AMCL

    •   Replacing the sensor with a more accurate one

    •   Adding sensors like IMU, visual odometry, and so on

Map Building

The map-building procedure is similar to that in the simulation process. The procedure is as follows:

1. Type any one of the following commands to perform the mapping according to your choice of input device:

    ```
    roslaunch bringup mapping.launch
    mode:=hardware device:=keyboard
    roslaunch bringup mapping.launch
    mode:=hardware device:=joystick
    roslaunch bringup mapping.launch
    mode:=hardware device:=android
    ```

2. Teleoperate the robot to create a map of the required area.

3. Save the map using the following command:

    ```
    rosrun map_server map_saver -f "map_name"
    ```

A sample map generated by Bumblebot in a building is shown in Figure 8-15.

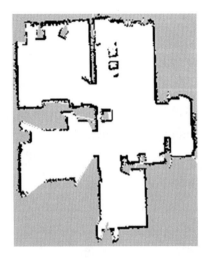

Figure 8-15. *Map created by Bumblebot in a building*

Autonomous Navigation

To perform autonomous navigation in Bumblebot, follow these steps:

1. Open a terminal.

2. Type the following command:

    ```
    roslaunch bringup navigation_hardware.launch
    ```

3. Give a goal position to the robot using the 2D Nav
    Goal button in RViz. See Figure 8-16.

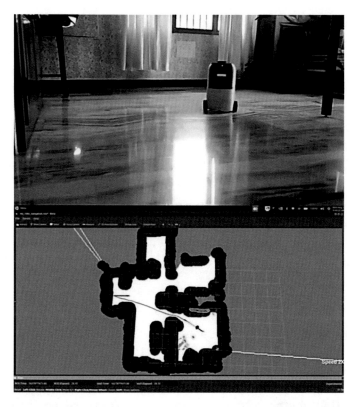

Figure 8-16. *Bumblebot performing navigation to the specified goal (top) and RViz visualization depicting the path, costmap, map, laser scan, and so on (bottom)*

Tuning the Navigation

Tuning the navigation involves adjusting the parameters of the navigation stack to improve the performance of a robot's navigation system. All the parameters can be tuned by adjusting the values in the `.yaml` files in the `BumbleBot_WS/src/bringup/config/` path.

Some of the frequently tuned parameters include these:

- **Inflation radius**: This parameter determines how far the robot should stay away from obstacles. If the radius is too small, the robot might collide with obstacles, while if it is too large, the robot might take a longer path to reach its goal.

- **Path distance bias**: This controls the balance between following the planned path closely and deviating from the path to avoid obstacles or other hazards. When the path distance bias is set to a high value, the navigation stack will prioritize staying close to the planned path even if it means the robot might collide with an obstacle. On the other hand, when the path distance bias is set to a low value, the navigation stack will prioritize deviating from the planned path to avoid obstacles or other hazards, even if it means the robot might take a longer path to reach its goal.

- **Goal distance bias**: This controls the balance between finding a path that is the shortest distance to the goal and finding a safer path. When the goal distance bias is set to a high value, the navigation stack will prioritize shorter paths even if they are riskier, which means that the robot might take unsafe shortcuts to reach its goal. On the other hand, when the goal distance bias is set to a low value, the navigation stack will prioritize safer paths even if they are longer, which means that the robot might take a longer path to avoid obstacles and ensure safe navigation.

- **Global planner parameters**: These parameters control the behavior of the global planner, which plans a path from the robot's current position to the goal position. Typically tuned parameters include the planning algorithm, distance threshold, and goal tolerance to optimize the path-planning process.

- **Local planner parameters**: These parameters control the behavior of the local planner, which adjusts the robot's velocity to follow the planned path. Typically tuned parameters include planning frequency, maximum velocity, and acceleration limits to ensure smooth and safe navigation.

- **Recovery behavior parameters**: These parameters control the robot's behavior when it encounters unexpected obstacles or errors. Typically tuned parameters include the number of recovery attempts and the distance threshold for detecting obstacles to ensure that the robot can recover from unexpected situations.

To easily modify the parameters in real time and visualize the effects, you can use the `rqt_reconfigure` tool. This tool can be started using the following command:

```
rosrun rqt_reconfigure rqt_reconfigure
```

This tool is shown in Figure 8-17. You can see the parameters of the local planner (here, it's DWA) within the `move_base` node that can be modified.

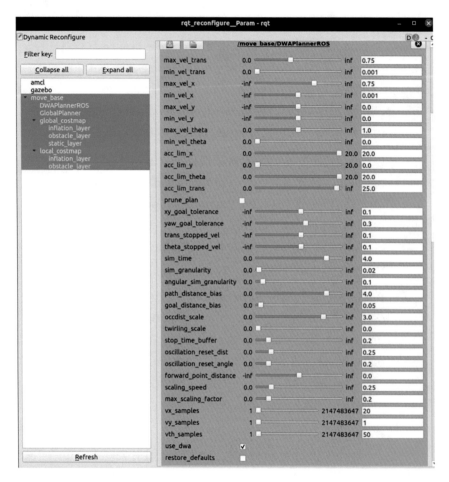

Figure 8-17. *The Rqt_Reconfigure tool*

The parameters you'll adjust depend on your robot's hardware, environment, and navigation requirements. It's important to carefully evaluate the performance of your robot's navigation after adjusting these parameters to ensure that the robot can navigate safely and efficiently.

For more information on navigation tuning, refer to the following links:

```
http://wiki.ros.org/navigation/Tutorials/Navigation%20
Tuning%20Guide
```

```
https://kaiyuzheng.me/documents/navguide.pdf
```

Robot Upstart

Robot Upstart is a ROS package that automatically starts a launch file when the robot is powered on. All the operations of Bumblebot are specified in a single launch file called automated_delivery.launch.

This launch file can be configured to start automatically when the robot is powered on using the robot_upstart package.

To install the robot_upstart package, use one of the following methods.

Method 1:

1. Open a terminal.

2. Enter this command:

   ```
   sudo apt-get install ros-noetic-
   robot-upstart
   ```

Method 2:

3. Open a terminal.

4. Enter these commands:

   ```
   cd BumbleBot_WS/src
   git clone https://github.com/
   clearpathrobotics/robot_upstart
   catkin_make
   ```

After installing the package, do the following:

1. Open a terminal.

2. Enter the following commands:

   ```
   rosrun robot_upstart install bringup/launch/
   automated_delivery/automated_delivery.launch
   --job bumblebot_start
   ```

```
sudo systemctl daemon-reload
sudo systemctl enable bumblebot_start
sudo reboot now
```

Where:

- `bumblebot_start` refers to the name of the service you are going to trigger during the startup of the robot.

- `bringup/launch/automated_delivery/automated_delivery.launch` refers to the launch file you want to start.

- The third command reloads the services.

- The fourth command enables your custom service called `bumblebot_start` to run during startup.

- The `reboot` command restarts the robot.

Note   If you encounter the `AttributeError: module 'enum' has no attribute 'IntFlag'` error, type this command in a terminal:

```
sudo pip uninstall enum34
```

From now on, whenever the robot is powered on, the robot will start its operations automatically.

Autonomous Delivery Application

At this point, you have the autonomous navigation properly set up and configured in your robot. Now you need to create a delivery application using the robot, so that you can command the robot whenever you want to deliver a payload from point A to point B. The flowchart for the delivery application is shown in Figure 8-18.

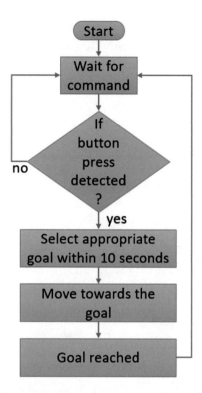

Figure 8-18. *Flowchart for autonomous delivery application*

In this example application, the robot is assumed to be at the home position and waits for a command. There are four preset goal locations on the map—home, goal-1, goal-2, and goal-3. The robot is configured in such a way that:

- If the pushbutton is pressed once, Home is selected.

- If the pushbutton is pressed twice, Goal-1 is selected.

- If the pushbutton is pressed three times, Goal-2 is selected.

- If the pushbutton is pressed four times, Goal-3 is selected.

After ten seconds from the last button press, the robot moves to the selected goal location. Once the robot reaches the goal location, it waits for the new goal.

Here's the code for the delivery application:

```python
#!/usr/bin/env python3

# Gets feedback from Move Base and triggers arduino functions

from datetime import datetime
import time
import rospy
from actionlib_msgs.msg import GoalStatusArray
from std_msgs.msg import String, Bool
from geometry_msgs.msg import PoseStamped

class DeliveryApplication:
    def push_button_callback(self, data):
        if data.data:  # Button Pressed
            if not self.first_press_detected:
                self.first_press_time = datetime.now()
                self.first_press_detected = True
            self.goal = self.goals[self.press_count]
            self.lcd_publisher.publish("")
            if self.press_count == 0:
                self.lcd_publisher.publish("Target: HOME")
            else:
                self.lcd_publisher.publish("Target: GOAL-" +
                str(self.press_count))
            self.press_count = (self.press_count + 1) %
            len(self.goals)
            time.sleep(1)
```

```python
    if self.first_press_detected:
        time_elapsed = (datetime.now() - self.first_press_
        time).total_seconds()
        if time_elapsed > 10:
            self.press_count = 0
            self.lcd_publisher.publish("Goal
            Selection    Complete")
            time.sleep(10)
            self.goal_publisher.publish(self.goal)
            self.first_press_detected = False

def callback(self, data):
    if not self.first_press_detected:
        if len(data.status_list) > 0:
            current_status = data.status_list[len
            (data.status_list)-1].status  # Get the
            status from the latest message
            print("Status of last goal: " + str
            (current_status))
            if current_status == 1:   # ACTIVE
                self.led_publisher.publish("yellow")
                self.lcd_publisher.publish("Goal
                Received")
                self.music_triggered = False
            elif current_status == 3:   # SUCCEEDED
                self.led_publisher.publish("green")
                self.lcd_publisher.publish("Goal
                Reached")
                time.sleep(1)
                if not self.music_triggered:
                    self.music_publisher.publish(True)
                    self.music_triggered = True
```

```
                        self.music_publisher.publish(False)
                else:   # OTHER
                    self.led_publisher.publish("red")
                    self.lcd_publisher.publish("Error...")

    def __init__(self):
        rospy.init_node('arduino_integration_node')

        # Goals
        self.home = PoseStamped()
        self.goal1 = PoseStamped()
        self.goal2 = PoseStamped()
        self.goal3 = PoseStamped()

        # Goal1
        self.goal1.header.frame_id = 'map'
        self.goal1.header.stamp = rospy.Time.now()
        self.goal1.pose.position.x = 7.80939388275
        self.goal1.pose.position.y = -2.87791061401
        self.goal1.pose.orientation.z = 0.999088019833
        self.goal1.pose.orientation.w = 0.0426981103338

        # Goal2
        self.goal2.header.frame_id = 'map'
        self.goal2.header.stamp = rospy.Time.now()
        self.goal2.pose.position.x = 4.8556022644
        self.goal2.pose.position.y = -6.80550956726
        self.goal2.pose.orientation.z = 0.717047815957
        self.goal2.pose.orientation.w = 0.697023980672

        # Goal3
        self.goal3.header.frame_id = 'map'
        self.goal3.header.stamp = rospy.Time.now()
        self.goal3.pose.position.x = 5.53890752792
```

```
self.goal3.pose.position.y = 0.579120635986
self.goal3.pose.orientation.z = -0.708675695394
self.goal3.pose.orientation.w = 0.705534378155

# Home
self.home.header.frame_id = 'map'
self.home.header.stamp = rospy.Time.now()
self.home.pose.position.x = 0.0399286746979
self.home.pose.position.y = -0.071713924408
self.home.pose.orientation.z = -0.00775012578072
self.home.pose.orientation.w = 0.999969967324

self.goals = [self.home, self.goal1, self.goal2,
self.goal3]

self.first_press_time = datetime.now()
self.press_count = 0
self.goal = self.home
self.music_triggered = False
self.current_status = 0  # Default
self.first_press_detected = False
self.led_publisher = rospy.Publisher('/led', String,
queue_size=10)
self.music_publisher = rospy.Publisher('/buzzer', Bool,
queue_size=10)
self.lcd_publisher = rospy.Publisher('/lcd', String,
queue_size=10)
rospy.Subscriber("/push_button", Bool, self.push_
button_callback, queue_size=1)
rospy.Subscriber("/move_base/status", GoalStatusArray,
self.callback, queue_size=1)
self.goal_publisher = rospy.Publisher('move_base_
simple/goal', PoseStamped, queue_size=10)
```

```
    # Performing Initial Operations
    time.sleep(5)
    self.led_publisher.publish("all")  # Blink all LEDs
    self.lcd_publisher.publish("Starting...")
    time.sleep(5)
    self.led_publisher.publish("none")  # Blink all LEDs
    self.lcd_publisher.publish("Ready")
    self.music_publisher.publish(False)
if __name__ == "__main__":
    delivery_app = DeliveryApplication()
    rospy.spin()
```

This is the Python script for the delivery application, which allows the user to select a goal location by pressing a button connected to the Arduino, and then sends the selected goal to the navigation stack. The script also updates the user with LED and LCD feedback about the current status of the robot and triggers a buzzer when the goal is reached.

Let's look in detail at the important parts of this code:

```
#!/usr/bin/env python3
```

This is called the "shebang" line, which lists the interpreter to be used to run the script. In this case, the script should run with the python3 interpreter.

```
from datetime import datetime
import time
import rospy
from actionlib_msgs.msg import GoalStatusArray
from std_msgs.msg import String, Bool
from geometry_msgs.msg import PoseStamped
```

These are `import` statements, which bring in required modules and message types:

- The `datetime` module works with dates and times.

- The `time` module pauses the program execution for a certain amount of time.

- The `rospy` module is the Python client library for ROS (Robot Operating System) and it writes ROS nodes in Python.

- `actionlib_msgs.msg` and `geometry_msgs.msg` are ROS message types that communicate with the `move_base` package, which provides the autonomous navigation capabilities of the robot.

  - The `GoalStatusArray` message provides feedback about the goal status (whether it succeeded, is active, was aborted, is pending, etc.).

  - The `PoseStamped` message provides goal locations to the robot.

- `std_msgs.msg` is a ROS message type used for general-purpose messaging. This example uses `String` and `Bool` messages to publish and subscribe topics to and from Arduino.

- `class DeliveryApplication` specifies a class called `DeliveryApplication` that contains several methods and data to call the autonomous delivery feature:

```
def push_button_callback(self, data):
    if data.data:  # Button Pressed
        if not self.first_press_detected:
            self.first_press_time = datetime.now()
```

```
            self.first_press_detected = True
        self.goal = self.goals[self.press_count]
        self.lcd_publisher.publish("")
        if self.press_count == 0:
            self.lcd_publisher.publish("Target: HOME")
        else:
            self.lcd_publisher.publish("Target: GOAL-"
            + str(self.press_count))
        self.press_count = (self.press_count + 1) %
        len(self.goals)
        time.sleep(1)

    if self.first_press_detected:
        time_elapsed = (datetime.now() - self.first_
        press_time).total_seconds()
        if time_elapsed > 10:
            self.press_count = 0
            self.lcd_publisher.publish("Goal Selection
            Complete")
            time.sleep(10)
            self.goal_publisher.publish(self.goal)
            self.first_press_detected = False
```

Here are some of the functions to note:

- push_button_callback is a callback function that is triggered when a message is received on the push_ button topic.

- The function takes one argument, called data, which is the message received on the topic.

- If data.data is True, it means that the button is pressed. The function sets the first_press_detected variable to True and assigns the current time to the first_press_time variable if it is the first button press.

- The corresponding goal from the goals list based on press_count is assigned to the goal variable.

- It then publishes a message to the /lcd topic to display the target goal.

- If the first_press_detected variable is True, it means that the first button press has already been detected.

- The function calculates the time elapsed since the first press and if it is greater than ten seconds, it resets the press_count variable to 0 and publishes a message to the /lcd topic to indicate that the goal selection is complete.

- It then publishes the selected goal to the /move_base_simple/goal topic and sets the first_press_detected variable to False.

```python
def callback(self, data):
    if not self.first_press_detected:
        if len(data.status_list) > 0:
            current_status = data.status_
            list[len(data.status_list)-1].
            status  # Get the status from the
            latest message
            print("Status of last goal: " +
            str(current_status))
        if current_status == 1:  # ACTIVE
```

```
                    self.led_publisher.
                    publish("yellow")
                    self.lcd_publisher.
                    publish("Goal Received")
                    self.music_triggered = False
                elif current_status == 3:
                # SUCCEEDED
                    self.led_publisher.
                    publish("green")
                    self.lcd_publisher.
                    publish("Goal Reached")
                    time.sleep(1)
                    if not self.music_triggered:
                        self.music_publisher.
                        publish(True)
                        self.music_triggered = True
                        self.music_publisher.
                        publish(False)
                else:  # OTHER
                    self.led_publisher.
                    publish("red")
                    self.lcd_publisher.
                    publish("Error...")
```

- The `callback` method is another callback function that is triggered every time a message is received on the `/move_base/status` topic.

- The function first checks if the `first_press_detected` variable is True or False. If it is True, it proceeds to process the message. Otherwise, it skips the message processing.

- The function then retrieves the status of the goal from the status_list field of the message.

- The status of the goal can be one of the following:

  1: Active

  3: Succeeded

  Other: Any other status

- If the status is 1 (ACTIVE), the function publishes a "yellow" color to the /led topic and Goal Received to the /lcd topic.

- It also sets the music_triggered variable to False.

- If the status is 3 (SUCCEEDED), the function publishes a "green" color to the /led topic and a Goal Reached status message to the /lcd topic.

- It then waits for one second and triggers the music by publishing True to the /buzzer topic.

- After playing the music once, it sets music_triggered to True and publishes False to the /buzzer topic to turn off the music.

- If the status is any other number, the function publishes a "red" color to the /led topic and sends an "Error..." status message to the /lcd topic.

```
def __init__(self):
    rospy.init_node('arduino_integration_node')

    # Goals
    self.home = PoseStamped()
    self.goal1 = PoseStamped()
    self.goal2 = PoseStamped()
```

```
self.goal3 = PoseStamped()

# Goal1
self.goal1.header.frame_id = 'map'
self.goal1.header.stamp = rospy.Time.now()
self.goal1.pose.position.x = 7.80939388275
self.goal1.pose.position.y = -2.87791061401
self.goal1.pose.orientation.z = 0.999088019833
self.goal1.pose.orientation.w = 0.0426981103338

# Goal2
self.goal2.header.frame_id = 'map'
self.goal2.header.stamp = rospy.Time.now()
self.goal2.pose.position.x = 4.8556022644
self.goal2.pose.position.y = -6.80550956726
self.goal2.pose.orientation.z = 0.717047815957
self.goal2.pose.orientation.w = 0.697023980672

# Goal3
self.goal3.header.frame_id = 'map'
self.goal3.header.stamp = rospy.Time.now()
self.goal3.pose.position.x = 5.53890752792
self.goal3.pose.position.y = 0.579120635986
self.goal3.pose.orientation.z = -0.708675695394
self.goal3.pose.orientation.w = 0.705534378155

# Home
self.home.header.frame_id = 'map'
self.home.header.stamp = rospy.Time.now()
self.home.pose.position.x = 0.0399286746979
self.home.pose.position.y = -0.071713924408
self.home.pose.orientation.z = -0.00775012578072
self.home.pose.orientation.w = 0.999969967324
```

```python
self.goals = [self.home, self.goal1, self.goal2,
self.goal3]

self.first_press_time = datetime.now()
self.press_count = 0
self.goal = self.home
self.music_triggered = False
self.current_status = 0  # Default
self.first_press_detected = False
self.led_publisher = rospy.Publisher('/led', String,
queue_size=10)
self.music_publisher = rospy.Publisher('/buzzer', Bool,
queue_size=10)
self.lcd_publisher = rospy.Publisher('/lcd', String,
queue_size=10)
rospy.Subscriber("/push_button", Bool, self.push_
button_callback, queue_size=1)
rospy.Subscriber("/move_base/status", GoalStatusArray,
self.callback, queue_size=1)
self.goal_publisher = rospy.Publisher('move_base_
simple/goal', PoseStamped, queue_size=10)

# Performing Initial Operations
time.sleep(5)
self.led_publisher.publish("all")  # Blink all LEDs
self.lcd_publisher.publish("Starting...")
time.sleep(5)
self.led_publisher.publish("none")  # Blink all LEDs
self.lcd_publisher.publish("Ready")
self.music_publisher.publish(False)
```

The __init__ method is the constructor that initializes various attributes of the class, creates publishers and subscribers, and sets up the ROS node for the application. Here is a brief explanation of what happens in the __init__ method:

- `rospy.init_node("delivery_application")`: Initializes the ROS node with the name delivery_ application.

- `self.goals = []`: Initializes an empty list to store the predefined delivery goals.

- `self.press_count = 0`: Initializes the button press count to 0.

- `self.first_press_detected = False`: Initializes a flag to indicate if the first button press has been detected.

- `self.first_press_time = None`: Initializes the timestamp of the first button press to None.

- `self.goal = None`: Initializes the current delivery goal to None.

- `self.music_triggered = False`: Initializes a flag to indicate if the music has been triggered after reaching a goal.

- `rospy.Subscriber("move_base/status", GoalStatusArray, self.callback)`: Creates a subscriber to the move_base/status topic with the GoalStatusArray message type and the self.callback function as the callback function.

- `rospy.Subscriber("push_button", Bool, self.push_button_callback)`: Creates a subscriber to the push_button topic with the Bool message type and the `self.push_button_callback` function as the callback function.

- `self.goal_publisher=rospy.Publisher("move_base_simple/goal",PoseStamped, queue_size=10)`: Creates a publisher to the move_base_simple/goal topic with the PoseStamped message type and a queue size of 10.

- `self.lcd_publisher = rospy.Publisher("lcd", String, queue_size=10)`: Creates a publisher to the lcd topic with the String message type and a queue size of 10.

- `self.led_publisher = rospy.Publisher("led", String, queue_size=10)`: Creates a publisher to the led topic with the String message type and a queue size of 10.

- `self.music_publisher = rospy.Publisher("music", Bool, queue_size=10)`: Creates a publisher to the music topic with the Bool message type and a queue size of 10.

```
# Performing Initial Operations
time.sleep(5)
self.led_publisher.publish("all")  # Blink all LEDs
self.lcd_publisher.publish("Starting...")
time.sleep(5)
self.led_publisher.publish("none")  # Blink all LEDs
self.lcd_publisher.publish("Ready")
self.music_publisher.publish(False)
```

This block of code in the constructor makes the robot wait for five seconds and then blinks all the LEDs and displays "Starting..." on the LCD screen for another five seconds. Then it turns off all the LEDs and displays "Ready" on the LCD screen. Finally, it publishes a False value to the music publisher to make sure that the music is not playing.

You must also define four predefined goal positions on the map as follows:

```
# Goals
self.home = PoseStamped()
self.goal1 = PoseStamped()
self.goal2 = PoseStamped()
self.goal3 = PoseStamped()

# Goal1
self.goal1.header.frame_id = 'map'
self.goal1.header.stamp = rospy.Time.now()
self.goal1.pose.position.x = 7.80939388275
self.goal1.pose.position.y = -2.87791061401
self.goal1.pose.orientation.z = 0.999088019833
self.goal1.pose.orientation.w = 0.0426981103338

# Goal2
self.goal2.header.frame_id = 'map'
self.goal2.header.stamp = rospy.Time.now()
self.goal2.pose.position.x = 4.8556022644
self.goal2.pose.position.y = -6.80550956726
self.goal2.pose.orientation.z = 0.717047815957
self.goal2.pose.orientation.w = 0.697023980672

# Goal3
self.goal3.header.frame_id = 'map'
self.goal3.header.stamp = rospy.Time.now()
self.goal3.pose.position.x = 5.53890752792
```

```
self.goal3.pose.position.y = 0.579120635986
self.goal3.pose.orientation.z = -0.708675695394
self.goal3.pose.orientation.w = 0.705534378155
```

- This code segment defines four variables of type PoseStamped—home, goal1, goal2, and goal3. The PoseStamped type represents the position and orientation of the goal locations.

- The code then defines the values for each of the four goals as follows:

  - The frame ID is set to map and the timestamp is set to the current time using rospy.Time.now().

  - The position and orientation values are then set for each goal using the pose.position and pose.orientation attributes, respectively.

Summary

This chapter explained the preliminary robot setup process, the Udev rules, configuring Arduino to control peripherals, 3D printing the robot parts, the electronics you need, assembling and wiring the robot, the motor gear ratio calculation process, the custom motor driver and ROS interface, the differential driver and odometry processes, teleoperation using ROS, odometry correction, including rotation and translation, building the map, autonomous navigation, tuning the navigation, starting the robot, and applying autonomous delivery. The next chapter explains how to add Lidar-based odometry generated by laser scan matching and how to add an inertial measurement unit (IMU) sensor. It also discusses how to perform sensor fusion to localize the robot more effectively.

# CHAPTER 9

Additional Sensors and Sensor Fusion

Outline

This chapter explains how to add lidar-based odometry generated by laser scan matching, and add an inertial measurement unit (IMU) sensor. It also discusses how to perform sensor fusion to optimize localizing the robot.

Odometry Sensors

Odometry plays a crucial role in estimating the position and orientation of a robot in the environment (i.e., localization). For a robot, several sources of odometry data can be used to estimate its motion. These sources provide information about the robot's movement, both linear and angular. Some of the common odometry sources for a robot are as follows:

1. **Wheel encoders**: Sensors mounted on the robot's wheels that measure the rotation or displacement of the wheels. By tracking the rotation or movement of the wheels, the wheel encoders can provide information about the robot's motion. Wheel encoders are commonly used in ground robots.

© Rajesh Subramanian 2023
R. Subramanian, *Build Autonomous Mobile Robot from Scratch using ROS*,
Maker Innovations Series, https://doi.org/10.1007/978-1-4842-9645-5_9

2. **Inertial Measurement Unit (IMU)**: An IMU typically consists of an accelerometer, a gyroscope, and sometimes a magnetometer. These sensors measure the following:

   - Accelerometer: Linear acceleration

   - Gyroscope: Angular velocity

   - Magnetometer: Magnetic field

   By integrating the accelerometer and gyroscope measurements over time, the IMU can provide information about the robot's linear and angular motion. IMUs are used to estimate rotation and sometimes velocity changes, but they can suffer from drift over time.

3. **GPS/GNSS**: Global Positioning System (GPS) or Global Navigation Satellite System (GNSS) receivers can provide accurate absolute position and velocity information. While GPS/GNSS can be used for long-term localization, they are less suitable for short-term odometry, due to their limited update rate and accuracy. GPS/GNSS can be used in combination with other odometry sources for localization.

4. **Visual odometry**: Visual odometry uses visual information from cameras to estimate motion. Visual odometry algorithms can estimate the robot's motion by analyzing the changes in images or video frames, the displacement of visual features, or the optical flow. Visual odometry can be used in combination with other odometry sources to improve accuracy.

5. **Lidar odometry**: Lidar sensors, which use laser beams to measure distances and generate 3D point clouds, can also be used for odometry estimation. Lidar odometry algorithms can estimate the robot's motion by comparing consecutive Lidar scans and tracking the movement of distinctive features or matching point clouds.

6. **Magnetic encoders**: Can be used as an alternative to wheel encoders for measuring the rotation of the robot's joints or components. They provide rotational displacement information that can be used for odometry estimation.

The combination and fusion of these odometry sources depend on the specific robot platform, its available sensors, and the desired accuracy and robustness of the odometry estimation. Integration techniques like sensor fusion (e.g., Kalman filters, particle filters) can be used to combine and optimize the information from multiple sources for more accurate motion estimation. The next section looks in detail at lidar-based odometry.

Lidar Based Odometry

As mentioned, there are various ways to obtain odometry data. Each method has its advantages and disadvantages. For instance, inertial measurement units (IMUs) are ideal for estimating spatial orientation but accumulate too much translational error over time. Odometry based on encoders has extensively been used to provide fast motion estimates for wheeled or legged robots. A drawback of this approach is that it is prone to be inaccurate due to wheel/leg slippage.

RF2O (Range Flow-based 2D Odometry) is a laser-based odometry algorithm designed for mobile robots. It estimates the motion of the robot by analyzing the range flow, which refers to the displacement of features between consecutive laser scans. The RF2O algorithm can be used in ROS with the rf2o_laser_odometry package (`https://github.com/MAPIRlab/rf2o_laser_odometry`). (Also, refer to `http://wiki.ros.org/rf2o_laser_odometry`).

Installation

To install the package, follow these steps:

1. Open a terminal (Ctrl+Shift+T).

2. Navigate to your robot workspace using the `cd` command. For example:
    ```
    cd BumbleBot_WS/
    ```

3. Go to the `src` folder:
    ```
    cd src
    ```

4. Clone the Git repository using this `git clone` command:
    ```
    git clone https://github.com/MAPIRlab/rf2o_laser_odometry.git
    ```

5. Check out the branch `ros1` for the `rf2o_laser_odometry` Git repository:
    ```
    git checkout ros1
    ```

6. Navigate to the parent directory as follows:
    ```
    cd ..
    ```

7. Compile the workspace using this command:
    ```
    catkin_make
    ```

Configuration

To configure the `rf2o_laser_odometry` ROS package, open the launch file and set these parameters:

```
<launch>

  <node pkg="rf2o_laser_odometry" type="rf2o_laser_odometry_
  node" name="rf2o_laser_odometry">
    <param name="laser_scan_topic" value="/scan"/>
    # topic where the lidar scans are being published
    <param name="odom_topic" value="/odom" />
    # topic where tu publish the odometry estimations
    <param name="publish_tf" value="true" />
    # whether or not to publish the tf::transform (base->odom)
    <param name="base_frame_id" value="base_footprint"/>
    # frame_id (tf) of the mobile robot base.
    A tf transform from the laser_frame to the base_frame is
    mandatory
    <param name="odom_frame_id" value="odom" />
    # frame_id (tf) to publish the odometry estimations
    <param name="init_pose_from_topic" value="" /> # (Odom
    topic) Leave empty to start at point (0,0)
    <!--param name="init_pose_from_topic" value="/base_pose_
    ground_truth" /--> # (Odom topic) Leave empty to start at
    point (0,0)

    <param name="freq" value="6.0"/>    #6
    # Execution frequency.
    <param name="verbose" value="true" />
    # verbose
  </node>

</launch>
```

This launch file contains many parameters:

- `<param name="laser_scan_topic" value="/scan"/>`:
  The ROS topic where the lidar scans are published.
  For Bumblebot, the lidar data is published under the
  `/scan` topic.

- `<param name="odom_topic" value="/odom" />`: This
  is the ROS topic where the odometry estimations are
  published. The default value is `/odom` and Bumblebot
  uses the same topic name.

- `<param name="publish_tf" value="true" />`: This
  determines whether the RF2O node should publish
  the transformation between `base_frame_id` and `odom_`
  `frame_id`. Set it to `true` to publish the transform, or to
  `false` to disable it. For Bumblebot, set it to `true`.

- `<param name="base_frame_id" value="base_`
  `footprint"/>`: This sets the frame ID (`tf`) of the mobile
  robot's base. It should match the actual frame ID of
  the robot's base. A `tf` transform from the `laser_frame`
  to the `base_frame_id` is mandatory for RF2O to work
  properly. For Bumblebot, set this parameter as `base_`
  `footprint`.

- `<param name="odom_frame_id" value="odom" />`:
  This sets the frame ID (`tf`) for publishing the odometry
  estimations. The default value is `odom`, and Bumblebot
  employs the same frame name.

- `<param name="init_pose_from_topic" value="" />`:
  This allows you to initialize the odometry estimation
  from an existing odometry topic. By default, it is left
  empty to start at the coordinate (0,0). If you want to

initialize the odometry from a specific topic (e.g.,
/base_pose_ground_truth), uncomment the line and
provide the appropriate topic name.

- `<param name="freq" value="6.0"/>`: This determines
  the execution frequency of the RF2O node, in Hertz.
  The default value is 6.0, but you can adjust it as needed.

- `<param name="verbose" value="true" />`: This sets
  whether the RF2O node should provide verbose output.
  If set to true, it will provide more detailed console
  output during execution.

Simulation Using Bumblebot

This section looks at simulating Bumblebot using laser odometry. For this
process, you need to configure the main launch file as well as the URDF
model file.

The following code shows how to configure Bumblebot launch's file for
the rf2o_laser_odometry package:

```xml
<?xml version="1.0"?>
<launch>

    <!-- Load the robot model into the parameter server -->
    <arg name="urdf_file" default="$(find xacro)/xacro --inorder
    '$(find my_robot_model)/urdf/my_robo_simulation_without_odom.
    urdf'"/>
    <param name="robot_description" command="$(arg urdf_file)" />

    <!-- Launch the gazebo world -->
    <include file="$(find gazebo_ros)/launch/empty_world.launch">
      <arg name="world_name" value="$(find my_robot_model)
      /gazebo_worlds/plaza_world.world"/>
```

```xml
</include>

<!-- Load the robot model in the parameter server into the
gazebo world -->
<node name="urdf_spawner" pkg="gazebo_ros" type="spawn_model"
respawn="false" output="screen" args="-urdf -model
my_robo -param robot_description"/>

<!-- Joint State Publisher - Publishes Joint Positions -->
<node name="joint_state_publisher" pkg="joint_state_
publisher" type="joint_state_publisher"/>

<!-- Robot State Publisher  - Uses URDF and Joint States to
compute Transforms -->
<node name="robot_state_publisher" pkg="robot_state_
publisher" type="robot_state_publisher"/>

<!-- RVIZ  - Visualization -->
<node name="rviz" pkg="rviz" type="rviz" args="-d $(find
my_robot_model)/rviz/my_robo_navigation.rviz"/>

<!-- Map server -->
<node pkg="map_server" name="map_server" type="map_server"
args="'$(find my_robot_model)/maps/plaza_world_map.yaml'"/>

<!-- Odom Node -->
 <include file="$(find rf2o_laser_odometry)/launch/rf2o_
 laser_odometry.launch"/>

<!-- AMCL - Localization -->
<node pkg="amcl" type="amcl" name="amcl" output="screen">
  <param name="odom_frame_id"              value="odom"/>
  <param name="base_frame_id"              value="base_
  footprint"/>
</node>

<!-- Move Base - Navigation -->
```

```
<node pkg="move_base" type="move_base" name="move_base"
output="screen">
    <rosparam file="$(find bringup)/config/costmap_common_
    params.yaml" command="load" ns="global_costmap"/>
    <rosparam file="$(find bringup)/config/costmap_common_
    params.yaml" command="load" ns="local_costmap"/>
    <rosparam file="$(find bringup)/config/local_costmap_
    params.yaml" command="load" />
    <rosparam file="$(find bringup)/config/global_costmap_
    params.yaml" command="load" />
    <rosparam file="$(find bringup)/config/global_planner_
    params.yaml" command="load" />

    <!-- GLOBAL PLANNERS -->
    <param name="base_global_planner" value="global_planner/
    GlobalPlanner"/>

    <!-- LOCAL PLANNERS -->
    <rosparam file="$(find bringup)/config/dwa_local_planner.
    yaml" command="load" />
    <param name="base_local_planner" value="dwa_local_planner/
    DWAPlannerROS"/>

    <!--rosparam file="$(find bringup)/config/trajectory_
    planner.yaml" command="load" />
    <param name="base_local_planner" value="base_local_planner/
    TrajectoryPlannerROS"/-->

    <!--rosparam file="$(find bringup)/config/teb_local_
    planner.yaml" command="load" />
    <param name="base_local_planner" value="teb_local_planner/
    TebLocalPlannerROS" /-->
</node>
</launch>
```

Let's look at the different sections of the launch file and their functionalities:

```
<?xml version="1.0"?>
```

This line specifies the XML version and initiates the launch file.

```
<arg name="urdf_file" default="$(find xacro)/xacro --inorder
'$(find my_robot_model)/urdf/my_robo_simulation_without_odom.
urdf'"/>
<param name="robot_description" command="$(arg urdf_file)" />
```

These lines define a urdf_file argument and a robot_description parameter that load the robot model into the ROS parameter server. The robot model is described in a URDF (Unified Robot Description Format) file, which is processed by the xacro tool to generate the actual URDF model. Adjust the file paths and names according to your specific robot model.

```
<include file="$(find gazebo_ros)/launch/empty_world.launch">
  <arg name="world_name" value="$(find my_robot_model)/gazebo_
  worlds/plaza_world.world"/>
</include>
```

This section launches the Gazebo simulation environment using the empty_world.launch file from the gazebo_ros package. It specifies the world file (plaza_world.world) that should be loaded. Adjust the file path and name to match the desired world.

```
<node name="urdf_spawner" pkg="gazebo_ros" type="spawn_model"
respawn="false" output="screen" args="-urdf -model
my_robo -param robot_description"/>
```

This node spawns the robot model in the Gazebo simulation using the spawn_model tool from the gazebo_ros package. It loads the URDF model from the robot_description parameter and sets the model name to my_ robo. Modify the model name as needed.

```
<node name="joint_state_publisher" pkg="joint_state_publisher"
type="joint_state_publisher"/>
```

This node publishes joint positions for visualization and debugging purposes using the joint_state_publisher package.

```
<node name="robot_state_publisher" pkg="robot_state_publisher"
type="robot_state_publisher"/>
```

This node computes the transforms between the robot's joints using the URDF model and publishes the transforms using the robot_state_ publisher package. This allows visualization tools, such as RViz, to display the robot in the correct pose.

```
<node name="rviz" pkg="rviz" type="rviz" args="-d $(find my_
robot_model)/rviz/my_robo_navigation.rviz"/>
```

This node launches RViz, a visualization tool, with a specific configuration file (my_robo_navigation.rviz). Adjust the file path and name to match the desired RViz configuration.

```
<node pkg="map_server" name="map_server" type="map_server"
args="'$(find my_robot_model)/maps/plaza_world_map.yaml'"/>
```

This node launches the map server, which loads a pre-generated map from a YAML file (plaza_world_map.yaml). Adjust the file path and name to match the desired map.

```
<include file="$(find rf2o_laser_odometry)/launch/rf2o_laser_
odometry.launch"/>
```

This line includes another launch file (rf2o_laser_odometry.launch) from the rf2o_laser_odometry package. It launches the RF2O laser odometry node along with its associated configurations. Make sure to have the rf2o_laser_odometry package properly installed and available in the ROS workspace.

```
<node pkg="amcl" type="amcl" name="amcl" output="screen">
  <param name="odom_frame_id" value="odom"/>
  <param name="base_frame_id" value="base_footprint"/>
</node>
```

This node launches the AMCL (Adaptive Monte Carlo Localization) package, which provides localization capabilities based on particle filters. It sets the frame ID of the odometry and base footprint according to the robot configuration.

```
<!-- Move Base - Navigation -->
<node pkg="move_base" type="move_base" name="move_base"
output="screen">
```

This node launches the Move Base package, which handles navigation tasks such as path planning and obstacle avoidance.

```
<rosparam file="$(find bringup)/config/costmap_common_params.
yaml" command="load" ns="global_costmap"/>
<rosparam file="$(find bringup)/config/costmap_common_params.
yaml" command="load" ns="local_costmap"/>
<rosparam file="$(find bringup)/config/local_costmap_params.
yaml" command="load" />
<rosparam file="$(find bringup)/config/global_costmap_params.
yaml" command="load" />
<rosparam file="$(find bringup)/config/global_planner_params.
yaml" command="load" />
```

These lines load various parameter files that configure the costmaps and planners used by Move Base. The `costmap_common_params.yaml` file is loaded twice, once for the global costmap (`ns="global_costmap"`) and once for the local costmap (`ns="local_costmap"`). These files define common parameters for both costmaps.

The `local_costmap_params.yaml` file contains parameters specific to the local costmap, which represents the immediate surroundings of the robot. The `global_costmap_params.yaml` file contains parameters for the global costmap, which covers a larger area.

The `global_planner_params.yaml` file contains parameters for the global planner, which plans the path for the robot from its current position to the goal position.

```
<!-- GLOBAL PLANNERS -->
<param name="base_global_planner" value="global_planner/
GlobalPlanner"/>
```

This line specifies the global planner to be used by Move Base. The `base_global_planner` parameter is set to `global_planner/GlobalPlanner`, indicating that the global planner of type `GlobalPlanner` is used. Adjust this parameter to use a different global planner.

```
<!-- LOCAL PLANNERS -->
<rosparam file="$(find bringup)/config/dwa_local_planner.yaml"
command="load" />
<param name="base_local_planner" value="dwa_local_planner/
DWAPlannerROS"/>
```

These lines load the parameters for the local planner, which handles local obstacle avoidance and trajectory generation for the robot. The `dwa_local_planner.yaml` file contains the specific parameters for the DWA (Dynamic Window Approach) local planner. The `base_local_planner` parameter is set to `dwa_local_planner/DWAPlannerROS`, indicating the use of the DWA local planner.

Now, let's look into configuring the URDF file of Bumblebot for the rf2o_laser_odometry package. The following code shows the differential drive plugin section of the URDF file.

```
<!--===== DIFF DRIVE PLUGIN - GAZEBO =====-->

<gazebo>
  <plugin name="differential_drive_controller"
  filename="libgazebo_ros_diff_drive.so">
      <!-- Plugin update rate in Hz -->
      <updateRate>10</updateRate>
      <!-- Name of left joint, defaults to `left_joint` -->
      <leftJoint>left_wheel_joint</leftJoint>
      <!-- Name of right joint, defaults to `right_joint` -->
      <rightJoint>right_wheel_joint</rightJoint>
      <!-- The distance from the center of one wheel to the
      other, in meters, defaults to 0.34 m -->
      <wheelSeparation>0.23</wheelSeparation>
      <!-- Diameter of the wheels, in meters, defaults to
      0.15 m -->
      <wheelDiameter>0.085</wheelDiameter>
      <!-- Wheel acceleration, in rad/s^2, defaults to 0.0
      rad/s^2 -->
      <wheelAcceleration>0.75</wheelAcceleration>
      <!-- Maximum torque which the wheels can produce, in Nm,
      defaults to 5 Nm -->
      <wheelTorque>500</wheelTorque>
      <!-- Topic to receive geometry_msgs/Twist message
      commands, defaults to `cmd_vel` -->
      <commandTopic>cmd_vel</commandTopic>
      <!-- Topic to publish nav_msgs/Odometry messages,
      defaults to `odom` -->
```

```
<odometryTopic>odom_unused</odometryTopic>
<!-- Odometry frame, defaults to `odom` -->
<odometryFrame>odom_unused</odometryFrame>
<!-- Robot frame to calculate odometry from, defaults to
`base_footprint` -->
<robotBaseFrame>base_footprint</robotBaseFrame>
<!-- Odometry source, 0 for ENCODER, 1 for WORLD,
defaults to WORLD -->
<odometrySource>1</odometrySource>
<!-- Set to true to publish transforms for the wheel
links, defaults to false -->
<publishWheelTF>false</publishWheelTF>
<!-- Set to true to publish transforms for the odometry,
defaults to true -->
<publishOdom>false</publishOdom>
<publishOdomTF>false</publishOdomTF>
<publishTF>false</publishTF>
<!-- Set to true to publish sensor_msgs/JointState on /
joint_states for the wheel joints, defaults to false -->
<publishWheelJointState>false</publishWheelJointState>
<!-- Set to true to swap right and left wheels, defaults
to true -->
<legacyMode>false</legacyMode>
  </plugin>
</gazebo>
```

You need to turn off the odometry and transformations published by Gazebo so that the navigation stack can use the odometry generated by the rf2o_laser_odometry package. You can do that by setting the differential drive plugin parameters of Gazebo as follows:

```
<!-- Set to true to publish transforms for the odometry,
defaults to true -->
<publishOdom>false</publishOdom>
<publishOdomTF>false</publishOdomTF>
<publishTF>false</publishTF>
```

Also, remap the odometry topic and odometry frame published by Gazebo to some other topic and frame name that's not subscribed by the navigation stack. This is done with the following code:

```
<!-- Topic to publish nav_msgs/Odometry messages, defaults to
`odom` -->
<odometryTopic>odom_unused</odometryTopic>
<!-- Odometry frame, defaults to `odom` -->
<odometryFrame>odom_unused</odometryFrame>
```

After configuring the files, you can run the simulation using the following command:

```
roslaunch bringup navigation_simulation_laser_odom.launch
```

Once the simulation has started, you can view the frames by using the following command:

```
rosrun tf2_tools view_frames.py
```

If everything is configured correctly, this will generate the frames as shown in Figure 9-1.

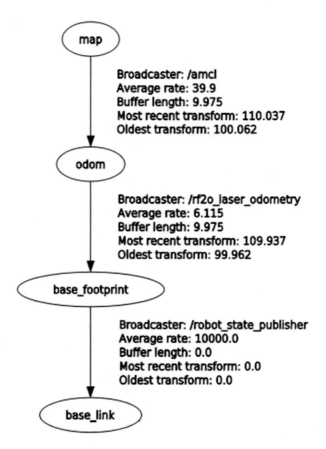

Figure 9-1. *TF frames diagram for Bumblebot using the rf2o_laser_ odometry package*

In Figure 9-1, you can see that the map frame, odom frame, and base_ footprint are connected. Also, the odom frame is published by the rf2o_ laser_odometry node. This means that the robot is navigating using the odometry generated by the rf2o_laser_odometry node.

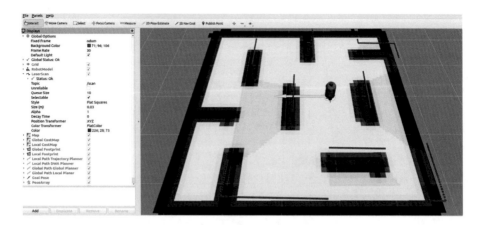

Figure 9-2. *RViz visualization and navigation using rf2o laser odometry*

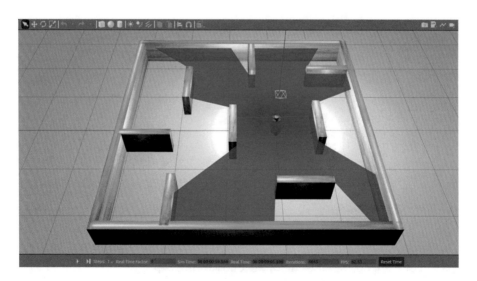

Figure 9-3. *Gazebo simulation and navigation using rf2o laser odometry*

Figures 9-2 and 9-3 illustrate the visualization and simulation for Bumblebot navigation using rf2o laser odometry.

IMU-Based Odometry

An IMU (Inertial Measurement Unit) is a sensor device that measures the linear and angular motion of an object. It typically combines multiple sensors to provide measurements of acceleration, angular velocity (gyroscope), and sometimes magnetic field (magnetometer) in a single package. The main components of an IMU are as follows:

- **Accelerometer**: Measures the acceleration experienced by the object along multiple axes. It provides information about linear motion and the force acting on the object.

- **Gyroscope**: Measures the angular velocity or rate of rotation around multiple axes. It provides information about the object's rotation or orientation changes.

- **Magnetometer**: Measures the strength and direction of the magnetic field. It provides information about the object's heading or orientation wrt the Earth's magnetic field.

Adding IMU to the Robot

There are two steps needed to add an IMU to the robot in simulation:

1. Add links and joints for the IMU in the URDF.

2. Add a Gazebo plugin for IMU in the URDF.

Adding Links and Joints for the IMU in the URDF

To add an IMU sensor to the robot, you need to create a link and joint in the URDF, as shown here:

```
<!--===== IMU LINK =====-->
<link name="imu_link">
  <inertial>
    <origin xyz="0 0 0" rpy="0 0 0" />
    <mass value="0.0093378" />
    <inertia ixx="0" ixy="0" ixz="0" iyy="0" iyz="0"
    izz="0" />
  </inertial>
  <visual>
    <origin xyz="0 0 0" rpy="0 0 0" />
    <geometry>
      <box size="0.05 0.01 0.02"/>
    </geometry>
    <material name="Red" />
  </visual>
  <collision>
    <origin xyz="0 0 0" rpy="0 0 0" />
    <geometry>
      <box size="0.05 0.01 0.02"/>
    </geometry>
  </collision>
</link>
<!--===== IMU JOINT =====-->
<joint name="imu_joint" type="fixed">
  <origin xyz="0.0075 0.0 0.25" rpy="0 0 3.14" />
  <parent link="base_link" />
  <child link="imu_link" />
  <axis xyz="0 0 0" />
</joint>
```

This code snippet describes the URDF (Unified Robot Description Format) representation of an IMU sensor in a robot model. Let's break down the code and look at its components:

- The `link` tag defines the IMU link in the robot model.

  - The `imu_link` tag specifies the name of the link.

  - The `inertial` tag describes the inertial properties (mass and inertia) of the IMU link. Here, the mass is set to 0.0093378 kg.

  - The `visual` tag defines the visual appearance of the IMU link. It uses a box geometry with the dimensions 0.05 x 0.01 x 0.02 meters with the color red.

  - The `collision` tag represents the collision properties of the IMU link. It uses the same box geometry as the visual representation.

- The `joint` tag defines the joint connecting the IMU link to the base link of the robot model.

  - The `imu_joint` tag specifies the name of the joint.

  - The `type` attribute is set to `fixed` since the IMU link is fixed wrt the base link.

  - The `origin` tag defines the position and orientation of the joint relative to the parent and child links. Here, it is set to a translation of (0.0075, 0.0, 0.25) meters and a rotation of (0, 0, 3.14) radians.

  - The `parent` tag specifies the parent link of the joint, which is set to `base_link`.

- The child tag specifies the child link of the joint, which is set to imu_link.

- The axis tag defines the axis of rotation for the joint. Here, it is set to (0, 0, 0), indicating no rotation.

Adding a Gazebo Plugin for IMU in the URDF

After the link and joint are created for the IMU in the URDF, you need to add the Gazebo plugin for the IMU sensor to the URDF file, as shown here:

```
<!--===== IMU PLUGIN - GAZEBO ======-->
<gazebo reference="imu_link">
  <gravity>true</gravity>
  <sensor name="imu_sensor" type="imu">
    <always_on>true</always_on>
    <update_rate>100</update_rate>
    <visualize>true</visualize>
    <topic>__default_topic__</topic>
    <plugin filename="libgazebo_ros_imu_sensor.so" name="imu_
    plugin">
      <topicName>imu_data</topicName>
      <bodyName>imu_link</bodyName>
      <updateRateHZ>10.0</updateRateHZ>
      <gaussianNoise>0.0</gaussianNoise>
      <xyzOffset>0 0 0</xyzOffset>
      <rpyOffset>0 0 0</rpyOffset>
      <frameName>imu_link</frameName>
      <initialOrientationAsReference>false</
      initialOrientationAsReference>
    </plugin>
    <pose>0 0 0 0 0 0</pose>
```

```
    </sensor>
</gazebo>
```

Let's break down the code and look at its components:

- The gazebo tag defines the Gazebo plugin for the IMU sensor.

- The reference attribute specifies the reference link for the plugin, which is set to imu_link.

- The sensor tag defines the properties of the IMU sensor.

- The name attribute sets the name of the sensor to imu_sensor.

- The type attribute specifies the type of the sensor, which is set to imu.

- The always_on tag indicates that the sensor is always active.

- The update_rate tag sets the update rate of the sensor to 100 Hz.

- The visualize tag enables visualization of the sensor in the Gazebo simulation.

- The plugin tag defines the IMU plugin to be used for the sensor.

- The filename attribute specifies the filename of the plugin, which is set to libgazebo_ros_imu_sensor.so.

- The name attribute sets the name of the plugin to imu_plugin.

- The `topicName` tag defines the ROS topic name on which the IMU data will be published, set to `imu_data`.

- The `bodyName` tag specifies the name of the body associated with the IMU sensor, set to `imu_link`.

- The `updateRateHZ` tag sets the update rate of the plugin to 10Hz.

- The `aussianNoise` tag specifies the amount of Gaussian noise to be added to the IMU data (set to 0.0 for no noise).

- The `xyzOffset` and `rpyOffset` tags define the translation and rotation offsets, respectively, for the IMU data.

- The `frameName` tag sets the name of the reference frame for the IMU sensor, set to `imu_link`.

- The `initialOrientationAsReference` tag indicates whether the initial orientation should be used as the reference (set to `false`).

- The `pose` tag specifies the initial pose of the IMU sensor. Here, it is set to (0, 0, 0, 0, 0, 0), indicating no initial translation or rotation.

Running the Simulation

To run the simulation, execute the following command in a terminal:

```
roslaunch bringup simulation_imu.launch
```

This will start the simulation of Bumblebot, including the IMU sensor. The visualization and simulation will look similar to that shown in Figures 9-4 and 9-5.

480

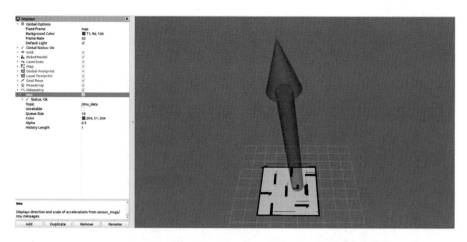

Figure 9-4. *RViz visualization of Bumblebot. The pink arrow indicates the IMU data (i.e., the direction and scale of accelerations)*

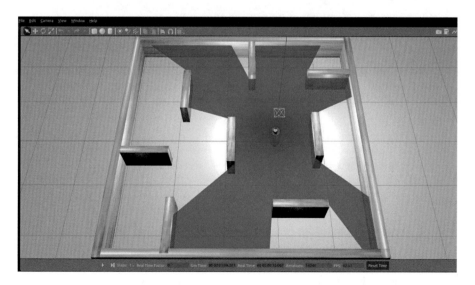

Figure 9-5. *Gazebo simulation of Bumblebot. The blue lines indicate the laser beams*

You can display the IMU data on the screen by using the following command:

```
rostopic echo /imu_data
```

This command will display the IMU data, as shown in Figure 9-6. You can see that the IMU messages contain orientation, angular velocity, and linear acceleration.

```
rajesh@ubuntu:~$ rostopic echo -n1 /imu_data
header:
  seq: 734
  stamp:
    secs: 129
    nsecs: 298000000
  frame_id: "imu_link"
orientation:
  x: 5.736190064646216e-06
  y: 2.2186690361384603e-05
  z: 0.9982607041041722
  w: 0.05895393215290318
orientation_covariance: [0.0, 0.0, 0.0, 0.0, 0.0, 0.0, 0.0, 0.0, 0.0]
angular_velocity:
  x: 0.027493859968900553
  y: -0.016161614972866337
  z: 0.001866522013281736
angular_velocity_covariance: [0.0, 0.0, 0.0, 0.0, 0.0, 0.0, 0.0, 0.0, 0.0]
linear_acceleration:
  x: 0.06641872057561882
  y: -0.20503704309797433
  z: 9.74903949326717
linear_acceleration_covariance: [0.0, 0.0, 0.0, 0.0, 0.0, 0.0, 0.0, 0.0, 0.0]
---
```

Figure 9-6. *Displaying IMU data*

The IMU message contains the following data:

- **Orientation**: The IMU message also holds estimates of the orientation or pose of the device relative to a reference frame. This is often be represented as a quaternion or as Euler angles (roll, pitch, and yaw).

- **Angular Velocity**: The IMU message contain angular velocity or the rate of rotation around the X, Y, and Z axes. It is commonly represented as a three-dimensional vector (angular velocity in radians per second).

- **Linear Acceleration**: The IMU message contains linear acceleration along the X, Y, and Z axes. It is represented as a three-dimensional vector (linear acceleration in meters per second squared).

Sensor Fusion

Sensor fusion is the process of combining data from multiple sensors to obtain a more accurate and comprehensive estimation of a system's state or environment. By fusing data from different sensors, the strengths of each sensor can be leveraged to compensate for the limitations and uncertainties of individual sensors, resulting in improved accuracy, robustness, and reliability.

In the context of robotics, sensor fusion plays a crucial role in perception, localization, and navigation. It allows the integration of data from sensors—such as cameras, Lidar, radar, GPS, IMUs, and more— to create a more complete and accurate representation of the robot's surroundings and its state.

There are different techniques and algorithms used for sensor fusion, including:

- **Kalman Filter**: The Kalman filter is a widely used algorithm for sensor fusion. It combines measurements from different sensors with the system's dynamic model to estimate the system's state while considering the uncertainties and noise associated with the sensor measurements.

- **Extended Kalman Filter (EKF)**: The EKF is an extension of the Kalman filter that allows nonlinear systems by linearizing the system model using a first-order approximation. It is commonly used for sensor fusion in systems with nonlinear dynamics.

- **Unscented Kalman Filter (UKF)**: The UKF is another extension of the Kalman filter that approximates the probability distribution of the system state using a set of sigma points. It provides more accurate estimates for systems with highly nonlinear dynamics.

- **Particle Filter (Monte Carlo Localization)**: The particle filter, also known as Monte Carlo Localization, is a probabilistic filtering technique that represents the system state using a set of particles. Each particle carries a hypothesis of the system state, and the filter updates and resamples the particles based on sensor measurements, allowing for robust estimation in environments with nonlinearities and uncertainty.

The choice of algorithm depends on the characteristics of the sensors, the application, and the desired level of accuracy and robustness.

In ROS, there are various libraries and packages available to perform sensor fusion, including:

- **Robot Localization** (the `robot_localization` package): Provides sensor fusion capabilities using extended and unscented Kalman filters for state estimation and sensor fusion in mobile robots.

- **Robot Pose EKF** (the `robot_pose_ekf` package): Combines odometry and IMU data to estimate the robot's pose using an extended Kalman filter.

- **Cartographer** (the `cartographer` package): A 2D and 3D simultaneous localization and mapping (SLAM) system that fuses data from multiple sensors, including Lidar and IMU, to create highly accurate maps of the environment.

In Bumblebot, you can use sensor fusion to combine IMU data and wheel odometry to get a more accurate and robust odometry estimate. The ROS package used for IMU-odometry fusion in this book is robot_pose_ekf. Refer to http://wiki.ros.org/robot_pose_ekf for more information.

Installation

To install the robot_pose_ekf package, follow these steps:

1. Open a terminal (Ctrl+Shift+T).

2. Navigate to your robot workspace using the cd command. For example:

    ```
    cd BumbleBot_WS/
    ```

3. Go to the src folder:

    ```
    cd src
    ```

4. Clone the Git repository using the git clone command:

    ```
    https://github.com/ros-planning/robot_
    pose_ekf.git
    ```

5. Navigate to the parent directory:

    ```
    cd ..
    ```

6. Compile the workspace using this command:

    ```
    catkin_make
    ```

Configuration

To configure the robot_pose_ekf package, open the launch file called robot_pose_ekf.launch, located in the BumbleBot_WS/src/robot_pose_ekf/launch folder. Then configure these parameters:

```
<launch>

<remap from="odom" to="/odom_diff_drive" />
<node pkg="robot_pose_ekf" type="robot_pose_ekf" name="robot_
pose_ekf">
  <param name="output_frame" value="odom"/>
  <param name="base_footprint_frame" value="base_footprint"/>
  <param name="freq" value="30.0"/>
  <param name="sensor_timeout" value="1.0"/>
  <param name="odom_used" value="true"/>
  <param name="imu_used" value="true"/>
  <param name="vo_used" value="false"/>
</node>

</launch>
```

The launch file is explained here:

- The <remap> tag redirects the /odom topic to a different topic, called /odom_diff_drive. The robot_pose_ekf node requires odometry input and subscribes to the /odom topic by default.

- The <node> tag runs the robot_pose_ekf node.

  - The pkg attribute is set to robot_pose_ekf, which specifies the package.

  - The type attribute specifies the executable to run, which is also set to robot_pose_ekf.

  - The name attribute specifies the name of the node, which is also set to robot_pose_ekf.

- The <param> tags set various parameters for the robot_pose_ekf node.

- The output_frame sets the output frame for the estimated robot pose, which is set to odom.

- The base_footprint_frame sets the frame associated with the robot's base footprint, which is set to base_footprint.

- The freq sets the frequency at which the node updates the estimated pose, which is set to 30.0Hz.

- The sensor_timeout sets the timeout value (in seconds) for sensor data reception. If no sensor data is received within this timeout, the node will stop updating the estimated pose.

- The odom_used specifies whether odometry data is used for sensor fusion, which is set to true.

- The imu_used specifies whether IMU data is used for sensor fusion, which is set to true.

- The vo_used specifies whether visual odometry data is used for sensor fusion, which is set to false.

After configuring the robot_pose_ekf launch file, you need to configure the differential drive plugin in the URDF file, as shown:

```
<!--===== DIFF DRIVE PLUGIN - GAZEBO =====-->

<gazebo>
  <plugin name="differential_drive_controller"
  filename="libgazebo_ros_diff_drive.so">
      <!-- Plugin update rate in Hz -->
      <updateRate>10</updateRate>
      <!-- Name of left joint, defaults to `left_joint` -->
      <leftJoint>left_wheel_joint</leftJoint>
      <!-- Name of right joint, defaults to `right_joint` -->
```

```
<rightJoint>right_wheel_joint</rightJoint>
<!-- The distance from the center of one wheel to the
other, in meters, defaults to 0.34 m -->
<wheelSeparation>0.23</wheelSeparation>
<!-- Diameter of the wheels, in meters, defaults to
0.15 m -->
<wheelDiameter>0.085</wheelDiameter>
<!-- Wheel acceleration, in rad/s^2, defaults to 0.0
rad/s^2 -->
<wheelAcceleration>0.75</wheelAcceleration>
<!-- Maximum torque which the wheels can produce, in Nm,
defaults to 5 Nm -->
<wheelTorque>500</wheelTorque>
<!-- Topic to receive geometry_msgs/Twist message
commands, defaults to `cmd_vel` -->
<commandTopic>cmd_vel</commandTopic>
<!-- Topic to publish nav_msgs/Odometry messages,
defaults to `odom` -->
<odometryTopic>odom_diff_drive</odometryTopic>
<!-- Odometry frame, defaults to `odom` -->
<odometryFrame>odom_diff_drive</odometryFrame>
<!-- Robot frame to calculate odometry from, defaults to
`base_footprint` -->
<robotBaseFrame>base_footprint</robotBaseFrame>
<!-- Odometry source, 0 for ENCODER, 1 for WORLD,
defaults to WORLD -->
<odometrySource>1</odometrySource>
<!-- Set to true to publish transforms for the wheel
links, defaults to false -->
<publishWheelTF>false</publishWheelTF>
```

```
<!-- Set to true to publish transforms for the odometry,
defaults to true -->
<publishOdom>true</publishOdom>
<publishOdomTF>false</publishOdomTF>
<!--publishTF>false</publishTF-->

<!-- Set to true to publish sensor_msgs/JointState on /
joint_states for the wheel joints, defaults to false -->
<publishWheelJointState>false</publishWheelJointState>
<!-- Set to true to swap right and left wheels, defaults
to true -->
<legacyMode>false</legacyMode>
    </plugin>
</gazebo>
```

The main things to notice here are as follows:

- Set odometryTopic to a different name other than the default one, say, odom_diff_drive. The default name is odom.

- Set odometryFrame to a different name other than the default one, say, odom_diff_drive. The default name is odom.

- Set publishOdom to true, so that the plugin computes and publishes odometry data.

- Set publishOdomTF to false, so that the plugin does not publish the transformation. This is to avoid conflict with the TF computed by the robot_pose_ekf node. The TF will be computed and published by the robot_pose_ekf node.

Simulation

To simulate autonomous navigation using odometry-IMU sensor fusion and Bumblebot, configure the launch file as follows:

```xml
<?xml version="1.0"?>
<launch>

  <!-- Load the robot model into the parameter server -->
  <arg name="urdf_file" default="$(find xacro)/xacro --inorder
  '$(find my_robot_model)/urdf/my_robo_simulation_imu.urdf'"/>
  <param name="robot_description" command="$(arg urdf_file)" />

  <!-- Launch the Gazebo world -->
  <include file="$(find gazebo_ros)/launch/empty_world.launch">
    <arg name="world_name" value="$(find my_robot_model)/
    gazebo_worlds/plaza_world.world"/>
  </include>

  <!-- Load the robot model in the parameter server into the
  gazebo world -->
  <node name="urdf_spawner" pkg="gazebo_ros" type="spawn_model"
  respawn="false" output="screen" args="-urdf -model
  my_robo -param robot_description"/>

  <!-- Joint State Publisher - Publishes Joint Positions -->
  <node name="joint_state_publisher" pkg="joint_state_
  publisher" type="joint_state_publisher"/>

  <!-- Robot State Publisher  - Uses URDF and Joint States to
  compute Transforms -->
  <node name="robot_state_publisher" pkg="robot_state_
  publisher" type="robot_state_publisher"/>

  <!-- RVIZ  - Visualization -->
```

```
<node name="rviz" pkg="rviz" type="rviz" args="-d $(find my_
robot_model)/rviz/my_robo_navigation.rviz"/>

<!-- Map server -->
<node pkg="map_server" name="map_server" type="map_server"
args="'$(find my_robot_model)/maps/plaza_world_map.yaml'"/>

<!-- Sensor Fusion: To combine odometry and IMU data -->
<include file="$(find robot_pose_ekf)/launch/robot_pose_ekf.
launch"/>

<!-- AMCL - Localization -->
<node pkg="amcl" type="amcl" name="amcl" output="screen">
    <rosparam file="$(find bringup)/config/amcl_fusion.yaml"
    command="load"/>
</node>

<!-- Move Base - Navigation -->
<node pkg="move_base" type="move_base" name="move_base"
output="screen">
  <rosparam file="$(find bringup)/config/costmap_common_
  params.yaml" command="load" ns="global_costmap"/>
  <rosparam file="$(find bringup)/config/costmap_common_
  params.yaml" command="load" ns="local_costmap"/>
  <rosparam file="$(find bringup)/config/local_costmap_
  params.yaml" command="load" />
  <rosparam file="$(find bringup)/config/global_costmap_
  params.yaml" command="load" />
  <rosparam file="$(find bringup)/config/global_planner_
  params.yaml" command="load" />

  <!-- GLOBAL PLANNERS -->
  <param name="base_global_planner" value="global_planner/
  GlobalPlanner"/>
```

```
<!-- LOCAL PLANNERS -->
<rosparam file="$(find bringup)/config/dwa_local_planner.
yaml" command="load" />
<param name="base_local_planner" value="dwa_local_planner/
DWAPlannerROS"/>

</node>

</launch>
```

The launch file is explained here:

- Load Robot Model

  - The arg tag defines an argument named urdf_
    file, which specifies the URDF file path of the
    robot model.

  - The param tag loads the URDF file into the
    parameter server using the robot_description
    parameter.

- Launch Gazebo World

  - The include tag launches the Gazebo simulation
    environment with an empty world.

  - The arg tag specifies the world file path.

- Spawn Robot Model in Gazebo

  - The node tag launches the spawn_model executable
    from the gazebo_ros package.

  - It is responsible for spawning the robot model into
    the Gazebo simulation environment.

  - The args attribute specifies the command-line
    arguments for the spawn_model executable,
    including the URDF model and robot description

- Joint State Publisher

  - It publishes the joint positions for the robot model.

  - The node tag launches the joint_state_publisher executable from the joint_state_publisher package.

- Robot State Publisher

  - It uses the URDF model and joint states to compute transformations.

  - The node tag launches the robot_state_publisher executable from the robot_state_publisher package.

- RViz Visualization

  - It is used for visualizing the robot model.

  - The node tag launches the rviz executable from the rviz package

  - The args attribute specifies the RViz configuration file

- Map Server

  - It provides a map from a YAML file

  - The node tag launches the map_server executable from the map_server package

- Sensor Fusion with robot_pose_ekf

  - It combines odometry and IMU data for sensor fusion

  - The include tag launches the robot_pose_ekf. launch file from the robot_pose_ekf package

- AMCL Localization

  - It performs localization

  - The node tag launches the amcl executable from the amcl package

  - The rosparam tag loads parameters from a YAML file

- Move Base Navigation

  - Handles navigation tasks, including global and local planning

  - The node tag launches the move_base executable from the move_base package

  - The rosparam tags load parameter files for configuring costmaps, planners, and other navigation settings

To start the simulation for autonomous navigation using odometry-sensor fusion in Bumblebot, run the following command in a terminal:

```
roslaunch bringup navigation_simulation_imu_fusion.launch
```

If everything goes well, you will get output similar to that shown in Figures 9-7 and 9-8 and you will be able to run autonomous navigation using odometry-IMU sensor fusion.

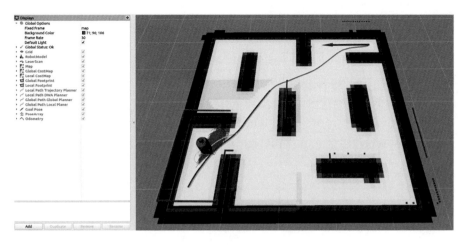

Figure 9-7. *RViz visualization of Bumblebot*

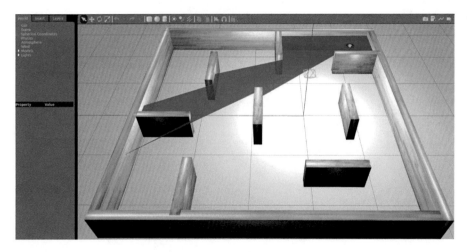

Figure 9-8. *Gazebo simulation of Bumblebot*

Summary

This chapter explained how to add Lidar-based odometry generated by laser scan matching, and how to add an inertial measurement unit (IMU) sensor. It also discussed how to perform sensor fusion to optimize localizing the robot more effectively. In the next chapter, you learn how to implement web interfacing and autodocking using ROS.

CHAPTER 10

Web Interface and Autodocking

Outline

This chapter explains how to implement web interfacing and autodocking using ROS.

Web Interface

A *web interface* is a graphical user interface (GUI) that is accessible through a web browser, enabling users to interact with and control a system or application over the Internet or a local network. Web interfaces are commonly used for a wide range of purposes, including controlling IoT devices, managing software applications remotely, accessing cloud services, and interacting with online platforms.

In the context of robotics and automation, web interfaces offer a convenient way to monitor and control robotic systems from various devices, such as laptops, tablets, and smartphones. They provide a user-friendly and accessible means of interacting with robots and viewing real-time data without the need for specialized software installations.

© Rajesh Subramanian 2023
R. Subramanian, *Build Autonomous Mobile Robot from Scratch using ROS*,
Maker Innovations Series, https://doi.org/10.1007/978-1-4842-9645-5_10

Some of the key components of a web interface include the following:

- **Frontend**: The frontend of the web interface is responsible for the user interface and user experience (UI/UX). It typically involves using web technologies like HTML, CSS, and JavaScript to create interactive elements, visual representations of data, and controls that users can interact with.

- **Backend**: The backend handles the communication with the robot or underlying system. In the case of a ROS-based robot, the backend interacts with the ROS system. Otherwise, the backend communicates with the robot's control system to send commands and receive data.

- **Web sockets or REST API**: To establish real-time communication between the web interface and the robot, web sockets or a RESTful API can be used. Web sockets allow bidirectional communication, making them ideal for real-time updates. On the other hand, a RESTful API is suitable for request-response type interactions.

- **Data visualization**: Data from the robot's sensors, state (e.g., waiting for a goal, moving to a goal, etc.), or any other relevant information can be displayed through texts, graphs, maps, or other visualizations on the web interface. This provides a clear overview of the robot's status and surroundings.

Web Interface in ROS

Creating a web interface in ROS involves integrating ROS with a web browser to allow users to interact with a robot system via a local network or the Internet. There is a package in ROS called ROSBridge that enables communication between ROS and web browsers using web sockets. ROSBridge is middleware that provides communication between ROS (Robot Operating System) and non-ROS environments, allowing data exchange and interaction with ROS from other systems, including web browsers, mobile applications, and other programming languages. It achieves this by translating ROS messages into a format that can be understood by non-ROS systems and vice versa.

ROSBridge is particularly useful when you want to create web interfaces and mobile apps or connect ROS to other technologies that don't natively support ROS communication. It facilitates the integration of ROS-powered robots and systems into a broader ecosystem, making it easier to interface with them from various platforms. Key features and components of ROSBridge include:

- **WebSockets**: ROSBridge primarily uses WebSockets as the communication protocol, enabling bidirectional real-time communication between ROS and web browsers. This allows web applications to subscribe to ROS topics, publish messages to ROS topics, and call ROS services.

- **JSON format**: To exchange data between ROS and non-ROS systems, ROSBridge converts ROS messages into JavaScript Object Notation (JSON) format, which is a widely used data interchange format. JSON makes it easy to represent complex data structures in a human-readable and lightweight manner.

- **Supported message types**: ROSBridge supports most
  ROS message types, allowing seamless communication
  of various data, such as sensor readings, robot state,
  control commands, and so on.

- **rosbridge_server**: The core component of ROSBridge
  is the rosbridge_server, which acts as a WebSocket
  server. It manages the WebSocket connections and
  handles the translation of ROS messages to and
  from JSON.

- **rosbridge_library**: The rosbridge_library
  is a Python library that provides the underlying
  functionality for the rosbridge_server. It offers the
  necessary tools to create custom ROSBridge-based
  applications if required.

Installing ROSBridge

To install ROSBridge, run the following command in a terminal:

```
sudo apt-get install ros-noetic-rosbridge-server
```

Building a Simple Web Page

This section shows you how to develop a simple web page with five
buttons and a textbox. The buttons correspond to five different predefined
goal locations. When a button is clicked, the robot is commanded to move
to the corresponding goal location autonomously. The textbox displays
the current status of the robot, such as "Waiting for Goal," "Goal Received,"
"Goal Reached," and "Error." The web page looks like Figure 10-1.

Figure 10-1. *Simple web page with five buttons and a textbox*

Here is the code for the web page:

```
<!DOCTYPE html>
<html>
<head>
    <title>Simple JavaScript Page</title>
    <style>
        body {
            display: flex;
            flex-direction: column;
            align-items: center;
            justify-content: center;
            height: 100vh;
```

```
    margin: 0;
    background-color: black; /* Set the background
    color to black */
}

.container {
    display: flex;
    flex-direction: column;
    align-items: center;
}

h1 {
    margin-bottom: 20px;
    color: white; /* Set the title text color to
    white */
}

button {
    margin: 5px;
}

/* Style for the textbox container */
.text-box-container {
    display: flex;
    align-items: center;
}

/* Style for the status label */
label {
    color: white;
    font-size: 16px;
    margin-right: 10px;
}
```

```
        /* Style for the textbox */
        input[type="text"] {
            width: 300px;
            padding: 10px;
            font-size: 16px;
        }
    </style>
</head>
<body>
    <div class="container">
        <h1>Bumblebot Commander</h1>

        <!-- Buttons for the goals -->
        <button onclick="publishMoveBaseGoal(1)">Goal 1
        </button>
        <button onclick="publishMoveBaseGoal(2)">Goal 2
        </button>
        <button onclick="publishMoveBaseGoal(3)">Goal 3
        </button>
        <button onclick="publishMoveBaseGoal(4)">Goal 4
        </button>
        <button onclick="publishMoveBaseGoal(5)">Home</button>

        <br><br>

        <!-- Textbox container with status label -->
        <div class="text-box-container">
            <label>Status:</label>
            <input type="text" id="textBox" placeholder=
            "Waiting for goal ...">
        </div>
    </div>
```

```
<!-- Add the roslib.js script -->
<script src="http://localhost:8000/ROS_Workspaces/ROS1/
BumbleBot_WS/src/roslibjs/build/roslib.min.js"></script>

<script>
    // Initialize ROS
    const ros = new ROSLIB.Ros({
        url: 'ws://localhost:9090' // Replace with your ROS
        bridge URL
    });

    const textBox = document.getElementById('textBox');
    // Update the textbox with initial status
    textBox.value = `Waiting for goal ...`;

    // Function to publish geometry_msgs/
    PoseStamped message
    function publishMoveBaseGoal(goalNumber) {
        const topic = new ROSLIB.Topic({
            ros: ros,
            name: '/move_base_simple/goal',
            messageType: 'geometry_msgs/PoseStamped'
        });

        const message = new ROSLIB.Message({
            header: {
                seq: 0,
                stamp: {
                    sec: 0,
                    nsec: 0
                },
                frame_id: 'map'
            },
```

```
pose: {
    position: {
        x: 0.0,
        y: 0.0,
        z: 0.0
    },
    orientation: {
        x: 0.0,
        y: 0.0,
        z: 0.0,
        w: 1.0
    }
}
});

// Update the goal coordinates based on the
goal number
switch (goalNumber) {
    case 1:
        message.pose.position.x = 2.518761396408081;
        message.pose.position.y = -1.5656793117523193;
        break;
    case 2:
        message.pose.position.x = -1.1461883783340454;
        message.pose.position.y = -1.8758769035339355;
        break;
    case 3:
        message.pose.position.x = -1.2489650249481201;
        message.pose.position.y = 2.0988407135009766;
        break;
```

```
        case 4:
            message.pose.position.x = 2.5786406993865967;
            message.pose.position.y = 2.017167329788208;
            break;
        case 5:
            message.pose.position.x = 0.0;
            message.pose.position.y = 0.0;
            break;
        default:
            console.log('Invalid goal number.');
            return;
    }

    topic.publish(message);
}

// Function to subscribe to /move_base/status topic
function subscribeToMoveBaseStatus() {
    const topic = new ROSLIB.Topic({
        ros: ros,
        name: '/move_base/status',
        messageType: 'actionlib_msgs/GoalStatusArray'
    });

    topic.subscribe(function(message) {
        if (message.status_list && message.status_list.
        length > 0) {
            // Get the status code of the latest status
            const statusCode = message.status_list
            [message.status_list.length - 1].status;

            // Display specific messages based on
            status code
```

```
            switch (statusCode) {
                case 1:
                    textBox.value = "Goal Received";
                    break;
                case 3:
                    textBox.value = "Goal Reached !";
                    break;
                default:
                    textBox.value = "Error...";
                    break;
            }
        }
    });
}

// Subscribe to /move_base/status topic
subscribeToMoveBaseStatus();
    </script>
</body>
</html>
```

The code contains the following parts:

- **HTML structure**: The HTML code defines the basic structure of the web page, including a container for the interface elements, a header displaying the title "Bumblebot Commander," buttons for different navigation goals, and a textbox that displays the status of the robot.

- **CSS styling**: CSS styling is used to format and style the elements on the web page. It sets the background color to black and the text color to white. It also centers the textbox, sets the font size, sets the margins, and so on.

- **JavaScript code:**

  - **ROS connection**: The JavaScript code uses the ROSlib library to establish a WebSocket connection with the ROS bridge server running at `ws://localhost:9090`. This allows bidirectional communication between the web interface and ROS.

  - **Buttons**: Each button in the web interface is associated with a specific navigation goal (Goal 1, Goal 2, Goal 3, Goal 4, and Home). When a button is clicked, the corresponding `publishMoveBaseGoal` function is called with the goal number as an argument.

  - **`publishMoveBaseGoal`**: This function creates a ROS topic named `/move_base_simple/goal` with the message type `geometry_msgs/PoseStamped`. It then publishes a goal message with predefined coordinates based on the provided goal input. The goal coordinates are updated in the `message.pose.position` object.

  - **`subscribeToMoveBaseStatus`**: This function subscribes to the `/move_base/status` topic to receive updates on the robot's navigation status. When a status message is received, the function checks the status code and updates the textbox with a corresponding status message based on the code. For example, `"Goal Received"` is displayed for status code 1, and `"Goal Reached!"` is displayed for status code 3.

- **ROSlib script**: The web page includes the ROSlib.js
  script, which is responsible for providing the necessary
  functionalities to establish the ROS connection and
  interact with ROS topics and messages.

Launch File

This section looks at the launch file that runs the simulation and web
interface.

```xml
<?xml version="1.0"?>
<launch>

  <!-- Load the robot model into the parameter server -->
  <param name="robot_description" command="$(find xacro)/
  xacro --inorder '$(find my_robot_model)/urdf/my_robo_
  simulation.urdf'"/>

  <!-- Launch the Gazebo world -->
  <include file="$(find gazebo_ros)/launch/empty_world.launch">
    <arg name="world_name" value="$(find my_robot_model)/
    gazebo_worlds/plaza_world.world"/>
  </include>

  <!-- Load the robot model in the parameter server into the
  Gazebo world -->
  <node name="urdf_spawner" pkg="gazebo_ros" type="spawn_model"
  respawn="false" output="screen" args="-urdf -model
  my_robo -param robot_description"/>

  <!-- Joint State Publisher - Publishes Joint Positions -->
  <node name="joint_state_publisher" pkg="joint_state_
  publisher" type="joint_state_publisher"/>
```

```xml
<!-- Robot State Publisher  - Uses URDF and Joint States to
compute Transforms -->
<node name="robot_state_publisher" pkg="robot_state_
publisher" type="robot_state_publisher"/>

<!-- Map server -->
<node pkg="map_server" name="map_server" type="map_server"
args="'$(find my_robot_model)/maps/plaza_world_map.yaml'"/>

<!-- AMCL - Localization -->
<node pkg="amcl" type="amcl" name="amcl" output="screen">
    <rosparam file="$(find bringup)/config/amcl.yaml"
    command="load"/>
</node>

<!-- Move Base - Navigation -->
<node pkg="move_base" type="move_base" name="move_base"
output="screen">
  <rosparam file="$(find bringup)/config/costmap_common_
  params.yaml" command="load" ns="global_costmap"/>
  <rosparam file="$(find bringup)/config/costmap_common_
  params.yaml" command="load" ns="local_costmap"/>
  <rosparam file="$(find bringup)/config/local_costmap_
  params.yaml" command="load" />
  <rosparam file="$(find bringup)/config/global_costmap_
  params.yaml" command="load" />
  <rosparam file="$(find bringup)/config/global_planner_
  params.yaml" command="load" />
  <rosparam file="$(find bringup)/config/move_base_params.
  yaml" command="load" />

  <!-- GLOBAL PLANNERS -->
  <param name="base_global_planner" value="global_planner/
  GlobalPlanner"/>
```

```
<!-- LOCAL PLANNERS -->
<rosparam file="$(find bringup)/config/dwa_local_planner.
yaml" command="load"/>
<param name="base_local_planner" value="dwa_local_planner/
DWAPlannerROS"/>
</node>

<!-- Http Server -->
<node name="http_server_node" pkg="web_interface"
type="start_http_server.sh" output="screen" />

<!-- Rosbridge Server -->
<node pkg="rosbridge_server" type="rosbridge_websocket"
name="rosbridge_server" output="screen">
  <param name="port" type="int" value="9090"/>
</node>

<!-- RVIZ  - Visualization -->
<node name="rviz" pkg="rviz" type="rviz" args="-d $(find my_
robot_model)/rviz/my_robo_navigation.rviz"/>

</launch>
```

This launch file does the following:

- **Loads the robot model**: The robot's URDF (Unified Robot Description Format) model is loaded into the parameter server using the param tag. It specifies the command to load the robot's URDF file using the xacro tool.

- **Launches the Gazebo world**: The launch file includes the empty_world.launch file from the gazebo_ros package, which launches Gazebo with an empty world. The value of the world_name argument is set to the path of the plaza_world.world file.

- **Spawns the robot model in Gazebo**: The urdf_ spawner node from the gazebo_ros package is used to spawn the robot model in Gazebo. It uses the robot_ description parameter to retrieve the robot's URDF and spawns it with the name my_robo.

- **Launches the joint state publisher**: The joint_state_ publisher node from the joint_state_publisher package is launched to publish joint positions for the robot.

- **Launches the robot state publisher**: The robot_ state_publisher node from the robot_state_ publisher package is launched to compute transforms between robot links based on the joint states and the URDF.

- **Launches the map server**: The map_server node from the map_server package is launched to provide a static map for localization and navigation. It loads the map file named plaza_world_map.yaml.

- **Launches the AMCL localization**: The amcl node from the amcl package is launched for localization. It localizes the robot using the map and sensor data. The AMCL parameters are loaded from the amcl.yaml file.

- **Launches the move base navigation**: The move_base node from the move_base package is launched for navigation. It handles global and local path planning. Various parameters for costmaps and planners are loaded from corresponding YAML files.

- **Launches the HTTP server**: A custom node called http_server_node is launched from the web_ interface package. It starts an HTTP server to serve the web interface files, allowing users to interact with the robot via a web browser.

- **Launches the ROSBridge server**: The rosbridge_ server node from the rosbridge_server package is launched to enable communication between ROS and web browsers through WebSockets. It sets the port to 9090, which is the default port for the ROSBridge server.

- **Launches the RViz visualization**: The rviz node from the rviz package is launched to visualize the robot's navigation. It loads a preconfigured RViz configuration file called my_robo_navigation.rviz.

Simulation with the Web Interface

To simulate Bumblebot with the web interface, type the following command:

```
roslaunch bringup
automated_delivery_simulation.launch
```

Now, open the Google Chrome web browser and type localhost:8000 in the address bar. This will open the web page that monitors and controls the robot.

These steps will produce output similar to that shown in Figure 10-2. You can now command the robot by providing goals and monitoring the status, as shown in Figures 10-3 and 10-4.

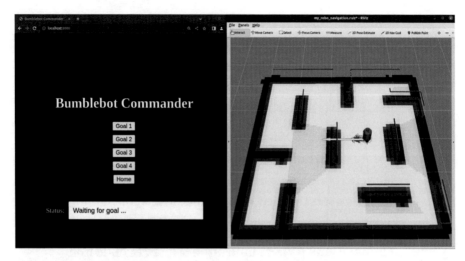

Figure 10-2. *Web interface for Bumblebot: waiting for a goal*

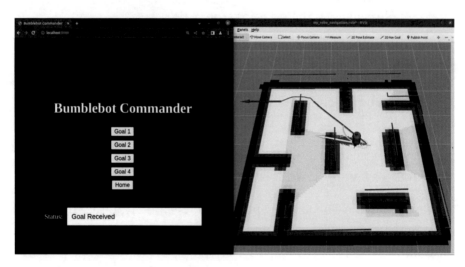

Figure 10-3. *Goal 1 button clicked*

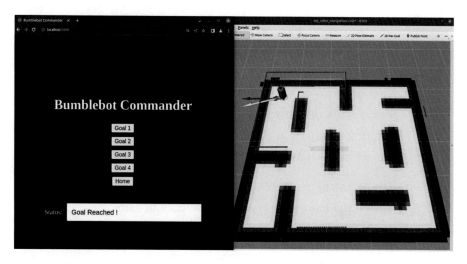

Figure 10-4. *Robot reached the selected destination*

Autodocking

This section explains another useful feature called *autodocking*. This feature enables the robot to autonomously find the charging station in the vicinity, navigate toward it, and dock, so that the battery can recharge. This is a valuable feature for robots that need to be able to operate for long periods without human intervention, as it eliminates the need for manual docking. The autodocking process typically involves several steps:

- **Detection**: The robot must be able to detect the presence and location of the docking station. This can be achieved using various sensors, such as cameras, Lidars, ultrasonic sensors, infrared sensors, and magnetic sensors.

- **Localization**: Once the robot detects the docking station, it needs to determine its position and orientation relative to the docking station. This is often accomplished using simultaneous localization and mapping (SLAM) algorithms or other localization techniques.

- **Navigation**: The robot plans a safe and efficient path to reach the docking station. It considers obstacles, dynamic changes in the environment, and other potential hazards during navigation.

- **Alignment**: The robot aligns itself with the docking station to ensure a successful connection. This might involve precise positioning and orientation adjustments.

- **Connection**: The robot completes the docking process by physically connecting itself to the charging or data transfer port on the docking station

- **Charging/transfer**: Once connected, the robot can start charging its batteries or perform the desired data transfers or communication tasks with the docking station.

Autodocking Using ROS

As mentioned, there are several ways to initiate autodocking in robots. This example uses an autodocking method that uses a camera and fiducial markers to perform autodocking. An existing autodocking package in ROS named `autodock` and provided by OSRF (Open-Source Robotics Foundation) is used to achieve autodocking capability. See `https://github.com/osrf/autodock` for more information.

The package uses a camera to detect ArUco markers placed on the charging station. ArUco stands for Augmented Reality University of Cordoba, where it was developed. An ArUco marker is a type of fiducial marker used in computer vision and augmented reality applications for visual tracking and pose estimation. ArUco markers are black-and-white square markers with a unique pattern of binary bits. They are designed to be easily detectable and identifiable by computer vision algorithms. This makes them ideal for applications that require accurate and robust marker detection in real time. The markers are typically printed on paper or displayed on a screen. An example of an ArUco marker is shown in Figure 10-5.

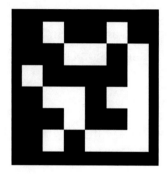

Figure 10-5. *An ArUco marker*

Installation

To install the autodock package, open a terminal and follow these steps.

1. Navigate into the src folder in the BumbleBot_WS workspace:

    cd BumbleBot_WS/src

2. Clone the autodock Git repository:

    git clone https://github.com/osrf/
    autodock.git

3. Navigate to the BumbleBot_WS folder:

    cd ..

4.   Install the required dependencies:

```
rosdep install --from-paths src --ignore-
src -r -y
```

5.   Compile and build the packages in the workspace:

```
catkin_make
```

Configuration

To configure the autodock package for Bumblebot, you need to set up
several files. The following subsections look into them one by one.

Configuring URDF

First, you need to add a camera to the kinematic description (URDF) of
Bumblebot. This is done as follows:

```
<!--====== CAMERA LINKS & JOINTS ======-->
  <joint name="camera_joint" type="fixed">
    <origin xyz="0.040 -0.011 0.130" rpy="0 -0.174 0"/>
    <parent link="base_link"/>
    <child link="camera_link"/>
  </joint>

  <link name="camera_link">
    <collision>
      <origin xyz="0.005 0.011 0.013" rpy="0 0 0"/>
      <geometry>
        <box size="0.015 0.030 0.027"/>
      </geometry>
    </collision>
  </link>
```

```
<joint name="camera_rgb_joint" type="fixed">
  <origin xyz="-0.153 0.011 0.009" rpy="0 0 3.14"/>
  <parent link="camera_link"/>
  <child link="camera_rgb_frame"/>
</joint>
<link name="camera_rgb_frame"/>

<joint name="camera_rgb_optical_joint" type="fixed">
  <origin xyz="0.0 0 0" rpy="-1.57 0 -1.57"/>
  <parent link="camera_rgb_frame"/>
  <child link="camera_rgb_optical_frame"/>
</joint>
<link name="camera_rgb_optical_frame"/>
```

Let's look at the different sections of this code:

- A fixed joint called camera_joint connects base_link to camera_link. The camera_link is represented as a box-shaped collision. It has a size of 0.015 meters in the x-direction, 0.030 meters in the y-direction, and 0.027 meters in the z-direction. The <origin> tag within the joint defines the position (xyz) and orientation (rpy) of camera_link relative to base_link.

- A fixed joint called camera_rgb_joint connects camera_link to camera_rgb_frame. The camera_rgb_frame is a link with no visual representation, as it lacks a <collision> or <visual> tag. The <origin> tag within the joint defines the position (xyz) and orientation (rpy) of the camera_rgb_frame relative to the camera_link.

- A fixed joint called camera_rgb_optical_joint connects
  camera_rgb_frame to camera_rgb_optical_frame.
  The camera_rgb_optical_frame is also a link with no
  visual representation. The <origin> tag within the joint
  defines the position (xyz) and orientation (rpy) of the
  camera_rgb_optical_frame relative to the camera_
  rgb_frame.

Next, you need to add a Gazebo plugin to simulate the camera. This is
done as follows:

```
<!--===== CAMERA PLUGIN - GAZEBO ======-->
<gazebo reference="camera_rgb_frame">
  <sensor type="camera" name="Pi Camera">
    <always_on>true</always_on>
    <visualize>true</visualize>
    <camera>
        <horizontal_fov>1.085595</horizontal_fov>
        <image>
            <width>320</width>
            <height>240</height>
            <format>R8G8B8</format>
        </image>
        <clip>
            <near>0.03</near>
            <far>100</far>
        </clip>
    </camera>
    <plugin name="camera_controller" filename="libgazebo_ros_
    camera.so">
      <alwaysOn>true</alwaysOn>
      <updateRate>30.0</updateRate>
      <cameraName>camera</cameraName>
```

```
      <frameName>camera_rgb_optical_frame</frameName>
      <imageTopicName>image</imageTopicName>
      <cameraInfoTopicName>camera_info</cameraInfoTopicName>
      <hackBaseline>0.07</hackBaseline>
      <distortionK1>0.0</distortionK1>
      <distortionK2>0.0</distortionK2>
      <distortionK3>0.0</distortionK3>
      <distortionT1>0.0</distortionT1>
      <distortionT2>0.0</distortionT2>
    </plugin>
  </sensor>
</gazebo>
```

Let's look at this code in detail:

- `<gazebo>`: This is the root tag for the Gazebo plugin configuration. It specifies that the plugin is attached to a model reference with the name `camera_rgb_frame`.

- `<sensor>`: Defines the camera sensor properties in Gazebo.

- `type="camera"`: Specifies that the sensor is a camera.

- `name="Pi Camera"`: The name assigned to this camera sensor.

- `<always_on>`: If set to `true`, the camera sensor is always active even if there's no subscriber. If set to `false`, the sensor is activated only when there's a subscriber to its data.

- `<visualize>`: If set to `true`, Gazebo will visualize the camera output, allowing you to see the camera's view in the simulation.

- `<horizontal_fov>`: Specifies the horizontal field of view of the camera (in radians).

- `<image>`: Describes the image properties.

- `<width>` and `<height>`: The width and height of the image produced by the camera.

- `<format>`: The format of the image. In this case, R8G8B8 refers to a standard 24-bit RGB color image.

- `<clip>`: Specifies the near and far clipping planes of the camera. Objects closer than the near value or farther than the far value will not be rendered in the camera feed.

- `<plugin>`: This tag indicates that a plugin is being added to the camera sensor.

- `name="camera_controller"`: The name of the camera plugin.

- `filename="libgazebo_ros_camera.so"`: The filename of the shared library (.so) containing the implementation of the camera plugin.

- `<alwaysOn>`: Indicates that the camera plugin is always running.

- `<updateRate>`: Specifies the update rate of the camera plugin in Hz (frames per second).

- `<cameraName>`: The name of the camera sensor (in this case, `camera`).

- `<frameName>`: The name of the frame to which the camera sensor is attached (`camera_rgb_ optical_frame`).

- `<imageTopicName>`: The topic name to publish the camera images (`image`).

- `<cameraInfoTopicName>`: The topic name to publish the camera calibration information (`camera_info`).

- `<hackBaseline>`: Specifies a hacky baseline for stereo rendering (usually set to 0.07).

- `<distortionK1>`, `<distortionK2>`, `<distortionK3>`, `<distortionT1>`, and `<distortionT2>`: Distortion coefficients for the camera (set to 0.0 in this case).

Here, the camera is attached to the rear side of the robot, because the charging port of Bumblebot is located on the backside.

Configuring Autodocking Parameters

You now need to create a configuration file to set the parameters for autodocking. The configuration file for Bumblebot is as follows:

```
# Gazebo Simulation robot


###############################################################
# Frame Ids

base_link : "base_footprint"
left_marker: "fiducial_10"
centre_marker: "fiducial_20"
right_marker: "fiducial_11"

###############################################################
# Vel profiles

# the upper lower limit of the cmd_vel output
linear_vel_range: [-0.5, 0.5]
angular_vel_range: [-0.13, 0.13]
```

```
# Absolute vel profile
# min (ang and linear) vel here means the min cmd_vel before
the robot stalls
max_linear_vel: 0.05        # m/s, for parallel.c and steer
min_linear_vel: 0.02        # m/s, for lastmile
max_angular_vel: 0.13       # rad/s
min_angular_vel: 0.05       # rad/s

##################################################################
# General Configs

cam_offset: 0.04            # camera to base link
stop_yaw_diff: 0.3 #0.03    # radian
stop_trans_diff: 0.02       # meters
tf_expiry: 1.0              # sec
controller_rate: 4.0        # hz
dock_timeout: 220           # sec
front_dock: false           # turtlebot "charging point" is
located in front

##################################################################
# Predock State

max_parallel_offset: 0.05 #0.005 #0.10    # m, will move to
                                          parallel.c if exceeded
predock_tf_samples: 5       # tf samples to avg, parallel.c
                             validation

##################################################################
# Steer Dock State

to_last_mile_dis: 0.5       # edge2edge distance where
                             transition to LM
to_last_mile_tol: 0.2       # transition tolerance from
                             SD to LM
```

```
# Determine how "rigorous" we want the robot to correct itself
during steer dock
offset_to_angular_vel: 2     # multiplier factor to convert
                               y-offset to ang vel

################################################################
# Last mile State

max_last_mile_odom: 0.20     # max last mile odom move without
                               using marker
stop_distance: 0.15 #0.1     # edge2edge distance to stop
                               from charger

################################################################
# Activate Charger

enable_charger_srv: False    # whether to activate charger after
                               last mile
check_battery_status: False # check if battery status is POWER_
                               SUPPLY_STATUS_CHARGING

################################################################
# Retry State

retry_count: 2               # how many times to retry
retry_retreat_dis: 0.5       # meters, distance retreat
                               during retry
```

The parameters are explained next:

- Frame IDs

  - base_link: The frame ID for the base link of
    the robot.

  - left_marker, center_marker, right_marker:
    Frame IDs for fiducial markers, used for
    localization or alignment during docking.

- Vel profiles:

  - `linear_vel_range`: Upper and lower limits of the linear velocity command output.

  - `angular_vel_range`: Upper and lower limits of the angular velocity command output.

  - `max_linear_vel`: Maximum linear velocity in meters per second (m/s).

  - `min_linear_vel`: Minimum linear velocity in m/s.

  - `max_angular_vel`: Maximum angular velocity in radians per second (rad/s).

  - `min_angular_vel`: Minimum angular velocity in rad/s.

- General configs:

  - `cam_offset`: Offset distance between the camera and the base link.

  - `stop_yaw_diff`: The maximum yaw difference (in radians) that triggers a stop during the docking process.

  - `stop_trans_diff`: The maximum translation difference (in meters) that triggers a stop during docking.

  - `tf_expiry`: The time (in seconds) after which a transformation (`tf`) is considered expired.

  - `controller_rate`: The rate (in Hz) at which the controller operates.

  - `dock_timeout`: Timeout (in seconds) for the docking process.

- `front_dock`: A Boolean value indicating whether the charging point is located in front of the robot. A `false` value indicates that the charging point is located on the rear side of the robot.

- Predock state:

  - `max_parallel_offset`: Maximum offset distance (in meters) to trigger a transition to another state if exceeded.

  - `predock_tf_samples`: Number of transformation (in `tf`) samples to average for validation.

- Steer dock state:

  - `to_last_mile_dis`: Edge-to-edge distance triggering a transition to the `last mile` state.

  - `to_last_mile_tol`: Tolerance for the transition from steer dock to the last mile.

  - `offset_to_angular_vel`: Multiplier factor to convert y-offset to angular velocity.

- Last mile state:

  - `max_last_mile_odom`: Maximum allowed odometry movement in the last mile state.

  - `stop_distance`: Edge-to-edge distance at which the robot stops from the charger.

- Activate charger:

  - `enable_charger_srv`: A Boolean value indicating whether to activate the charger after the last mile.

  - `check_battery_status`: A Boolean value indicating whether to check the battery status before activating the charger.

- Retry state:

  - retry_count: Number of times to retry the docking.

  - retry_retreat_dis: Distance (meters) to retreat during the retry process.

Note that the configuration file for Bumblebot is located at the following path:

```
BumbleBot_WS/src/autodocking/autodock_examples/configs/
bumblebot.yaml
```

Configuring the Launch File

Here is the main launch file for autodocking with autonomous navigation:

```
<?xml version="1.0"?>
<launch>

  <!-- Global param -->
  <arg name="autodock_server" default="true"/>
  <arg name="use_sim_time" default="true"/>
  <arg name="debug_mode" default="true" doc="Use this to have
  aruco detections always ON (publish fiducial images always)
  and other useful visualization markers. Turn OFF for better
  performance in deployment."/>

  <!-- Load the robot model into the parameter server -->
  <param name="robot_description" command="$(find xacro)/
  xacro --inorder '$(find my_robot_model)/urdf/my_robo_
  simulation_with_camera.urdf'"/>

  <!-- Launch the gazebo world -->
  <include file="$(find gazebo_ros)/launch/empty_world.launch">
    <arg name="world_name" value="$(find my_robot_model)/
    gazebo_worlds/plaza_world.world"/>
```

```
<arg name="use_sim_time" value="$(arg use_sim_time)"/>
<arg name="gui" value="true"/>
<arg name="headless" value="false"/>
<arg name="debug" value="false"/>
<env name="GAZEBO_MODEL_PATH" value="$(find autodock_sim)/
models" />
</include>

<!-- Load the robot model in the parameter server into the
gazebo world -->
<node name="urdf_spawner" pkg="gazebo_ros" type="spawn_model"
respawn="false" output="screen" args="-urdf -model
my_robo -param robot_description"/>

<!-- Joint State Publisher - Publishes Joint Positions -->
<node name="joint_state_publisher" pkg="joint_state_
publisher" type="joint_state_publisher"/>

<!-- Robot State Publisher  - Uses URDF and Joint States to
compute Transforms -->
<node name="robot_state_publisher" pkg="robot_state_
publisher" type="robot_state_publisher"/>

<!-- Map server -->
<node pkg="map_server" name="map_server" type="map_server"
args="'$(find my_robot_model)/maps/plaza_world_map.yaml'"/>

<!-- AMCL - Localization -->
<node pkg="amcl" type="amcl" name="amcl" output="screen">
    <rosparam file="$(find bringup)/config/amcl.yaml"
    command="load"/>
</node>
```

```xml
<!-- Move Base - Navigation -->
<node pkg="move_base" type="move_base" name="move_base"
output="screen">
  <rosparam file="$(find bringup)/config/costmap_common_
  params.yaml" command="load" ns="global_costmap"/>
  <rosparam file="$(find bringup)/config/costmap_common_
  params.yaml" command="load" ns="local_costmap"/>
  <rosparam file="$(find bringup)/config/local_costmap_
  params.yaml" command="load" />
  <rosparam file="$(find bringup)/config/global_costmap_
  params.yaml" command="load" />
  <rosparam file="$(find bringup)/config/global_planner_
  params.yaml" command="load" />
  <rosparam file="$(find bringup)/config/move_base_params.
  yaml" command="load" />

  <!-- GLOBAL PLANNERS -->
  <param name="base_global_planner" value="global_planner/
  GlobalPlanner"/>

  <!-- LOCAL PLANNERS -->
  <rosparam file="$(find bringup)/config/dwa_local_planner.
  yaml" command="load"/>
  <param name="base_local_planner" value="dwa_local_planner/
  DWAPlannerROS"/>
</node>

<!-- Http Server -->
<node name="http_server_node" pkg="web_interface"
type="start_http_server.sh" output="screen" />

<!-- Rosbridge Server -->
```

```
<node pkg="rosbridge_server" type="rosbridge_websocket"
name="rosbridge_server" output="screen">
  <!-- Set the port for the websocket server (default:
  9090) -->
  <param name="port" type="int" value="9090"/>
</node>

<!-- RVIZ  - Visualization -->
<node name="rviz" pkg="rviz" type="rviz" args="-d $(find my_
robot_model)/rviz/my_robo_autodocking.rviz"/>

<!--===== AutoDocking Stuff =====-->

<!-- Launch AutoDock Server Node -->
<group if="$(arg autodock_server)">
  <include file="$(find autodock_core)/launch/autodock_
  server.launch">
    <arg name="autodock_config" default="$(find autodock_
    examples)/configs/bumblebot.yaml"/>
    <!--arg name="autodock_config" default="$(find autodock_
    examples)/configs/mock_robot.yaml"/-->
    <arg name="debug_mode" value="$(arg debug_mode)" />
  </include>
</group>

<!-- launch fiducial detect  -->
<node pkg="aruco_detect" name="aruco_detect"
      type="aruco_detect" output="log" respawn="false">
  <param name="/use_sim_time" value="$(arg use_sim_time)"/>
  <param name="image_transport" value="compressed"/>
  <param name="publish_images" value="true" />
  <param name="dictionary" value="8"/>
  <param name="do_pose_estimation" value="true"/>
```

```
    <param name="verbose" value="false"/>
    <param name="fiducial_len_override" value="10: 0.1,
    11: 0.1, 20: 0.04"/>
    <remap from="camera/compressed" to="camera/image/
    compressed"/>
    <remap from="camera_info" to="camera/camera_info"/>
  </node>

  <!-- obstacle observer node  -->
  <node type="obstacle_observer"
        name="obstacle_observer" pkg="autodock_core"
        output="screen">
    <param name="vicinity_radius" value="0.16"/>
    <param name="coverage_percent" value="0.35"/>
    <param name="occupancy_prob" value="70"/>
    <param name="base_link_name" value="base_link"/>
    <remap from="/move_base/global_costmap/local_costmap" to="/
move_base/local_costmap/costmap"/>
  </node>

  <!-- spawn the mini mock charger -->
  <node pkg="gazebo_ros" type="spawn_model" name="spawn_model"
        args="-sdf -database MiniMockCharger -model
        charger -x -0.5" />

  <!-- Autodock Trigger -->
  <node pkg="custom_scripts" name="autodock_trigger"
type="autodock_trigger.py"/>

</launch>
```

The provided XML file is a launch file that's used to launch a Gazebo simulation for a robot with autodocking capabilities. It sets up the robot's URDF description, loads the Gazebo world, and launches various nodes for localization, navigation, visualization, and autodocking functionalities.

Here are the main sections of the launch file:

- Global parameters

  - `autodock_server`: A Boolean argument that determines whether to launch the `autodock` server or not.

  - `use_sim_time`: A Boolean argument that determines whether to use simulated time.

  - `debug_mode`: A Boolean argument to enable/disable additional visualization markers for debugging.

- Load robot model

  - The robot model (URDF) is loaded into the parameter server using xacro to parse the URDF file and generate the `robot_description` parameter.

- Launch the Gazebo world

  - The Gazebo world is launched using the `empty_world.launch` file, which sets up the Gazebo environment.

  - The `GAZEBO_MODEL_PATH` environment variable specifies the path to the custom models used in the simulation.

- URDF spawner

  - The robot model is spawned in the Gazebo world using the `spawn_model` node from `gazebo_ros`.

- Joint state publisher and robot state publisher

  - These nodes are used to publish joint positions and compute transforms based on the URDF and joint states.

- Map server

  - The `map_server` node is launched to serve a map for localization.

- AMCL (Adaptive Monte Carlo Localization)

  - The `amcl` node is launched for localization using Adaptive Monte Carlo Localization.

- Move base

  - The `move_base` node is launched for navigation using the Move Base package.

  - Various ROS parameter files are loaded to configure the costmaps and planners for global and local navigation.

- HTTP server

  - A custom node (`start_http_server.sh`) is launched for an HTTP server.

- ROSBridge server

  - The `rosbridge_server` node is launched for bridging ROS communication with a WebSocket.

- RViz visualization

  - RViz is launched with a custom configuration file (`my_robo_autodocking.rviz`) for visualization.

- Autodocking setup

  - The autodock server node (`autodock_server. launch`) is launched to handle the autodocking behavior.

  - Fiducial detection using the `aruco_detect` node is set up to detect ArUco markers for localization.

  - An obstacle observer node (`obstacle_observer`) is launched to observe obstacles in the robot's vicinity.

  - Spawn mini mock charger

  - A mock charger is spawned in Gazebo using the `gazebo_ros spawn_model` node.

  - Autodock trigger

  - A custom node (`autodock_trigger.py`) is launched to trigger the autodocking behavior.

You can now manually trigger autodocking by issuing the following command in a terminal:

```
rostopic pub /autodock_action/goal autodock_core/
AutoDockingActionGoal {} --once
```

Optionally, we can trigger the autodocking by issuing another custom command in a terminal:

```
rostopic pub /start_autodock std_msgs/Empty "{}"
```

Configuring the Web Interface

This simple web page contains several buttons that enable you to control the robot. In addition to the goal buttons, you can add one more buttons to trigger the autodock. The web page looks like Figure 10-6.

Figure 10-6. *Web page with Autodock trigger button*

The code for this web page is as follows:

```
<!DOCTYPE html>
<html>
<head>
    <title>Bumblebot Commander</title>
    <style>
        body {
            display: flex;
            flex-direction: column;
            align-items: center;
            justify-content: center;
            height: 100vh;
            margin: 0;
```

```css
    background-color: black; /* Set the background
    color to black */
}

.container {
    display: flex;
    flex-direction: column;
    align-items: center;
}

h1 {
    margin-bottom: 20px;
    color: white; /* Set the title text color to
    white */
}

button {
    margin: 5px;
}

/* Style for the textbox container */
.text-box-container {
    display: flex;
    align-items: center;
}

/* Style for the status label */
label {
    color: white;
    font-size: 16px;
    margin-right: 10px;
}

/* Style for the textbox */
```

```
        input[type="text"] {
            width: 300px;
            padding: 10px;
            font-size: 16px;
        }
    </style>
</head>
<body>
    <div class="container">
        <h1>Bumblebot Commander</h1>

        <!-- Buttons for the goals -->
        <button onclick="publishMoveBaseGoal(1)">Goal 1
        </button>
        <button onclick="publishMoveBaseGoal(2)">Goal 2
        </button>
        <button onclick="publishMoveBaseGoal(3)">Goal 3
        </button>
        <button onclick="publishMoveBaseGoal(4)">Goal 4
        </button>
        <button onclick="publishMoveBaseGoal(5)">Home</button>
        <button onclick="publishAutodockGoal()">Autodock
        </button>
        <br><br>

        <!-- Textbox container with status label -->
        <div class="text-box-container">
            <label>Status:</label>
            <input type="text" id="textBox" placeholder="Enter
            text here">
        </div>
    </div>
```

```html
<!-- Add the roslib.js script -->
<script src="http://localhost:8000/roslibjs/build_roslibjs/
roslib.min.js"></script>
<script>
    // Initialize ROS
    const ros = new ROSLIB.Ros({
        url: 'ws://localhost:9090' // Replace with your
        ROSBridge URL
    });

    // JavaScript code remains the same
    const textBox = document.getElementById('textBox');
    // Update the textbox with initial status
    textBox.value = `Waiting for goal ...`;

    // Function to publish geometry_msgs/
    PoseStamped message
    function publishMoveBaseGoal(goalNumber) {
        const topic = new ROSLIB.Topic({
            ros: ros,
            name: '/move_base_simple/goal',
            messageType: 'geometry_msgs/PoseStamped'
        });

        const message = new ROSLIB.Message({
            header: {
                seq: 0,
                stamp: {
                    sec: 0,
                    nsec: 0
                },
                frame_id: 'map'
            },
```

```
            pose: {
                position: {
                    x: 0.0,
                    y: 0.0,
                    z: 0.0
                },
                orientation: {
                    x: 0.0,
                    y: 0.0,
                    z: 0.0,
                    w: 1.0
                }
            }
        });

        // Update the goal coordinates based on the
        goal number
        switch (goalNumber) {
            case 1:
                message.pose.position.x = 2.518761396408081;
                message.pose.position.y = -1.5656793117523193;
                break;
            case 2:
                message.pose.position.x = -1.1461883783340454;
                message.pose.position.y = -1.8758769035339355;
                break;
            case 3:
                message.pose.position.x = -1.2489650249481201;
                message.pose.position.y = 2.0988407135009766;
                break;
```

```
        case 4:
            message.pose.position.x = 2.5786406993865967;
            message.pose.position.y = 2.017167329788208;
            break;
        case 5:
            message.pose.position.x = 0.0;
            message.pose.position.y = 0.0;
            break;
        default:
            console.log('Invalid goal number.');
            return;
    }

    topic.publish(message);
}

// Function to publish the Autodock goal to /start_
autodock topic
function publishAutodockGoal() {
    const textBox = document.getElementById('textBox');
    textBox.value = "Starting autodock sequence ...";
    const topic = new ROSLIB.Topic({
        ros: ros,
        name: '/start_autodock',
        messageType: 'std_msgs/Empty'
    });

    const message = new ROSLIB.Message({});
    topic.publish(message);
}
```

```
// Function to subscribe to /move_base/status topic
function subscribeToMoveBaseStatus() {
    const topic = new ROSLIB.Topic({
        ros: ros,
        name: '/move_base/status',
        messageType: 'actionlib_msgs/GoalStatusArray'
    });

    topic.subscribe(function(message) {
        if (message.status_list && message.status_list.
        length > 0) {
            // Get the status code of the latest status
            const statusCode = message.status_
            list[message.status_list.length -
            1].status;

            // Display specific messages based on
            status code
            switch (statusCode) {
                case 1:
                    textBox.value = "Goal Received";
                    break;
                case 3:
                    textBox.value = "Goal Reached !";
                    break;
                default:
                    textBox.value = "Error...";
                    break;
            }
        }
    });
}
```

```
    // Subscribe to /move_base/status topic
    subscribeToMoveBaseStatus();

  </script>
</body>
</html>
```

This HTML file provides a user-friendly web interface to interact with Bumblebot. It allows the user to send predefined navigation goals and trigger the autodocking sequence. The robot's status during navigation is displayed as text in real time on the web page. Let's dig into various parts of this code:

- HTML structure

    - The HTML file begins with the usual HTML structure, including `<html>`, `<head>`, and `<body>` tags.

- Styling

    - The `<style>` section contains CSS styling rules for formatting the UI components.

    - The background color is set to black, and various elements are styled for proper alignment and appearance.

- Container and title

    - The main content is placed in a container `<div>` element with the `container` class.

    - The "Bumblebot Commander" title is displayed as an `<h1>` heading in the container.

- Goal buttons

  - Six buttons are provided for selecting predefined navigation goals for the robot: Goal 1, Goal 2, Goal 3, Goal 4, Home, and Autodock.

  - Each button has an associated onclick attribute that calls JavaScript functions to publish the corresponding goal to the robot.

- Textbox container and status label

  - Below the buttons, there is a textbox container with the text-box-container class.

  - Inside this container, there is a <label> element with the Status: text that indicates the purpose of the textbox.

- Textbox for status

  - The textbox for displaying the robot's status has the <input> tag with type="text".

  - It has the ID textBox, which is later used to update its content dynamically using JavaScript.

- JavaScript script

  - The JavaScript code is embedded in the <script> section.

  - It first initializes a ROS connection by creating a ROS object with the URL of the ROSBridge server.

  - The publishMoveBaseGoal(goalNumber) function publishes navigation goals to the /move_base_ simple/goal topic based on the selected goal. The goal coordinates are updated accordingly.

- The publishAutodockGoal() function publishes a message to trigger autodocking on the /start_ autodock topic when the Autodock button is clicked.

- The subscribeToMoveBaseStatus() function subscribes to the /move_base/status topic to receive updates about the robot's navigation status. Based on the received status, it updates the textbox with relevant messages, such as "Goal Received", "Goal Reached!" or "Error...".

Simulation

To simulate autodocking along with autonomous navigation and the web interface, run the following command in a terminal:

```
roslaunch bringup autodocking_sim.launch
```

This will open the RViz and Gazebo windows, as shown in Figures 10-7 and 10-8, respectively. You can see the charging station shown as a white box with three ArUco markers behind the robot, as shown in Figure 10-8.

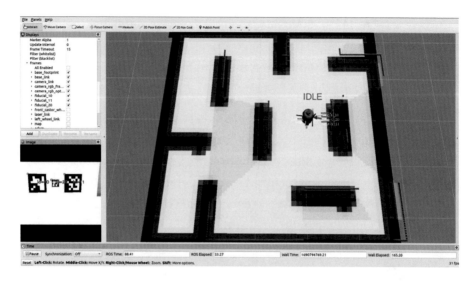

Figure 10-7. *Visualization for Bumblebot with autodocking capability*

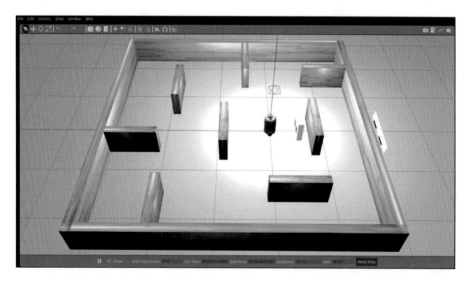

Figure 10-8. *Simulation for Bumblebot with autodocking capability*

Open the Google Chrome browser and type the following in the
address bar:

```
localhost:8000
```

This will open the web page to the Bumblebot Commander, as shown
in Figure 10-9.

Figure 10-9. *Web page to command Bumblebot with autodocking
capability*

You can now command the robot to move to the desired goal positions
by clicking the Goal buttons and you can initiate autodocking using the
Autodock button. Before commanding the robot to perform autodocking,
make sure that the robot is back in the Home position. The robot requires
the ArUco markers within its vicinity. Figures 10-10 and 10-11 illustrate
the screenshots when the autodock sequence has been triggered and
accomplished, respectively.

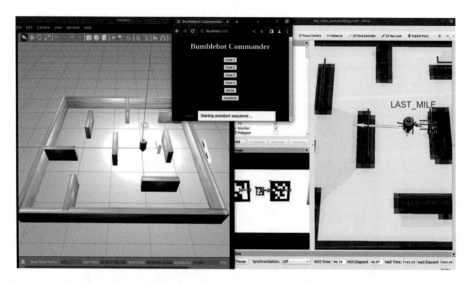

Figure 10-10. *Autodock sequence triggered after clicking the Autodock button*

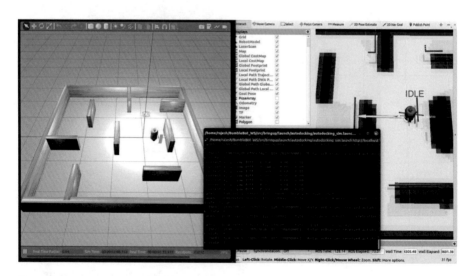

Figure 10-11. *Autodocking completed*

Summary

This chapter explained how to implement a web interface to monitor and command various functionalities of a robot and how to perform autodocking using ROS.

Index

A

© Rajesh Subramanian 2023
R. Subramanian, *Build Autonomous Mobile Robot from Scratch using ROS*,
Maker Innovations Series, https://doi.org/10.1007/978-1-4842-9645-5

F

G

H